THE GREAT SHIFT

Lexicons of Survival: Exploring the Anthropocene through Emerging
Terms and Trends in Climate Fiction.

INSIGHT PUBLICA®

Nadakkavu, Kozhikode, Kerala, 673011
www.insightpublica.com
e-mail: insightpublica@gmail.com
Title: THE GREAT SHIFT
Author: Anver Kabeer, Arya V. Sathyan
First Edition: May 2025

Printed and Published by
InsightinPublica Printers & Publishers Pvt. Ltd.
ISBN: 978-93-5517-999-9
₹250/-

THE GREAT SHIFT

Lexicons of Survival: Exploring the Anthropocene through Emerging Terms and Trends in Climate Fiction.

ANVER KABEER, ARYA V. SATHYAN

INSIGHT PUBLICA

THE GREAT SHIFT

Lexicons of Survival: Exploring the Anthropocene
through Emerging
Terms and Trends in Climate Fiction.

12 Cli-fi terms and 35 + authors.

"For The Almighty, Our Loving Parents and Family"

- AK and AVS

CONTENTS

FOREWORD 11

PREFACE 15

ACKNOWLEDGEMENTS 17

INTRODUCTION 19

CHAPTER 1 23

CHAPTER 2 36

CHAPTER 3 44

CHAPTER 4 53

CHAPTER 5 66

CHAPTER 6 73

CHAPTER 7 85

CHAPTER 8 93

CHAPTER 9 102

CHAPTER 10 116

CHAPTER 11 132

CHAPTER 12 147

CONCLUSION 155

REFERENCE 158

FOREWORD

Dr. Mathew Dan

As humanity stands at the crossroads of ecological collapse, social upheaval, and rapid technological transformation, we find ourselves in urgent need of new narratives—stories that not only reflect the world as it is, but challenge us to imagine the world as it could be. In this context, climate fiction—or cli-fi—has emerged as more than just a literary subgenre; it represents a powerful cultural response to the pressing realities of the Anthropocene. Through speculative storytelling, cli-fi captures the anxieties, hopes, and contradictions of our era, offering both a mirror to our times and a window into possible futures. Considering the increasing environmental crises around the world, there is a growing call for a 'green shift'—an urgent emphasis on the conservation of nature and its finite resources. It is in this scenario that the book The Great Shift: Lexicons of Survival: Exploring the Anthropocene through Emerging Terms and Trends in Climate Fiction attains special relevance.

This book serves as both a scholarly resource and a literary map for those seeking to understand the significance of climate fiction in contemporary thought. Drawing on a wide array of sources and texts, it presents a well-curated exploration of twelve foundational elements that shape the cli-fi genre—ranging from landmark authors and canonical works to core concepts and critical themes. The result is a cohesive and accessible book that will be invaluable to educators, students, researchers, and general readers interested in environmental humanities, speculative fiction, and the broader intersections of culture and climate.

The authors navigate the genre's diverse terrain with clarity and insight. Their analysis reveals not only the aesthetic richness of cli-fi, but also its capacity to engage deeply with ecological ethics, *climate justice*, posthumanism, and resilience. Each chapter invites readers to think critically about the stories we tell in the face of crisis—and how those stories, in turn, shape our responses to ecological and existential threats.

One of the great strengths of this book lies in its interdisciplinary approach. The authors have artfully interwoven literary criticism, environmental philosophy, science studies, and cultural analysis to provide a nuanced understanding of climate fiction's role in contemporary discourse. By doing so, they position cli-fi not merely as a form of entertainment or speculation, but as a vital intellectual tool—one that can illuminate the challenges of the present while imagining alternative futures.

In a world increasingly defined by uncertainty and transformation, cli-fi offers a space for reflection, resistance, rethinking, and reinvention. This book captures that spirit with both precision and passion, resulting in a meaningful and timely contribution. Whether you are a seasoned academic or a newcomer to the genre, you will find in these pages a thoughtful and engaging companion to the literature of our warming world. The ruthless interventions in nature by humans are no longer mere fiction. This book is a call to reimagine our relationship with the Earth, with each other, and with the stories that shape our collective future. I congratulate the authors on bringing together the profound insights of various writers and thematically organizing them into a volume that reads like a crown beautifully embedded with jewels.

(Dr. Mathew Dan, Former Principal Scientist at JNTBGRI, Palode, is a veteran botanist credited with discovering seven new species and honoured with four awards and an internship at the Royal Botanic Gardens, Kew, UK.)

PREFACE

In recent years, climate fiction—or (Cli-Fi)—has emerged as a compelling literary genre that reflects the urgent realities of our time. As humanity faces the escalating threats of climate change, ecological collapse, and sustainability crises, (Cli-Fi) provides a narrative space where the entanglements between human societies, technological advances, and the natural world are critically examined.

This book engages with that narrative space, offering readers a comprehensive exploration of the key concepts, authors, and texts that define the genre. Drawing on a range of literary works—from the dystopian vision of Margaret Atwood's *Oryx and Crake* to the philosophical insights of *Timothy Morton's Dark Ecology*—this volume captures the evolution and diversity of climate fiction. It explores how literature can act as both a mirror and a guide, reflecting our anxieties about environmental degradation while also offering imaginative frameworks for resilience, adaptation, and hope.

At the heart of this study are twelve essential concepts—such as *Marsification, Greenwashing, and Deep Adaptation*—that serve as entry points into the genre's critical discourse. By unpacking these terms and linking them to representative texts, the book fosters a deeper appreciation for how (Cli-Fi) interrogates the interconnected systems shaping life on Earth.

Intended for students, academics, environmentalists, and general readers alike, this work bridges disciplines to illuminate the vital role literature plays in responding to the climate crisis. Whether you are delving into (Cli-Fi) for scholarly inquiry or personal reflection, this guide invites you to consider how stories about ecological futures can reshape our understanding of the present.

As we continue to navigate the uncertainties of the Anthropocene, climate fiction offers not just warnings, but visions—ways of seeing, thinking, and imagining that are essential to fostering environmental consciousness and action. This book aspires to contribute meaningfully to that endeavour by mapping the landscape of (Cli-Fi) and underscoring its relevance to our collective future.

Anver Kabeer
Arya V. Sathyan

ACKNOWLEDGEMENTS

We extend our heartfelt gratitude to Insight Publica, Kozhikode, for believing in our vision and bringing this book to life. Our sincere thanks to Dr. S Hari Krishnan for his creative contribution in designing the cover page, which beautifully captures the essence of our work.

We have been fortunate to receive invaluable mentorship from Dr. Manoj Chandrasenan and Vithura V. Asok, whose guidance and editorial support played a pivotal role in shaping this book. Their encouragement and critical insights were instrumental throughout our writing journey.

Anver Kabeer wishes to express his deepest gratitude to his lifelong mentor, Sri P. Vijayan IPS, for his timely mentoring and for inspiring a positive outlook that has influenced every stage of his life. He is profoundly thankful to Lovely Professional University, Phagwara, Punjab, for nurturing his research and writing skills. Special thanks go to his research guide, Dr. Rayees Ahmad Bhat, for his steadfast support and guidance.

Anver also extends heartfelt thanks to his family, especially his father J. M. Kabeer and mother Nujoomath Beevi, for their unconditional love and blessings. His heartfelt appreciation goes to his wife, Shibina S. Thaham, and children, Abdul Basith S. A. and Khadheeja Fathima S. A., whose love and patience were his constant source of strength. He is also indebted to his siblings, Ansar Kabeer and Aneesa N., for their encouragement. Anver gratefully acknowledges his colleagues at Govt. V&HSS, Vithura, and the entire Student Police Cadet Programme team for providing a vibrant space that fostered his growth in research, social work, training, and writing.

Arya V. Sathyan conveys her sincere appreciation to the University of Technology, Jaipur, for providing a fertile academic environment that supported her research journey. She is especially thankful to her research guide, Dr. Ravi Yadav, for his insightful mentorship and unwavering encouragement.

This book is not just an academic milestone but also a cherished dream realized—a dream once held close by her father and uplifted by her mother's prayers. Arya offers her gratitude to her parents, K. Sathyadevan, a retired army official, and C. Valsala, for their love, support, and blessings that sustained her every step of the way.

She also thanks her husband, V. Sujithkumar, a revenue official, whose strength and constant support have been the cornerstone of her accomplishments. Her daughters, Dia A. Sujith and Dhaya A. Sujith, have been her joy, inspiration, and greatest motivators throughout this journey. Arya is grateful to her aunt, C. Sujatha, and sister, Aiswarya V. S., for their unwavering belief in her. She extends her gratitude to her friends, colleagues, and the management of Lekshmi Vilasom High School, Pothencode, Thiruvananthapuram, for their continuous encouragement and support.

This book stands not only as a reflection of our individual efforts, but as a testament to the love, belief, and support of all those who have walked beside us in this journey.

INTRODUCTION

THE SHIFT IN STORYTELLING

From Stereotypes to Survival: Reimagining Literature for a Changing World

For centuries, storytelling has revolved around universal human themes—love, war, heroism, identity—that resonate across cultures and eras. These narratives, while timeless, often mirrored an idealized, static vision of the world, relying on predictable stereotypes and individual struggles to craft compelling tales. But as humanity enters the Anthropocene—a geological epoch defined by human impact on the planet—the stories we tell are undergoing a profound transformation. With climate crises, social upheavals, and the challenges of survival at the forefront of global consciousness, literature is evolving to reflect the urgency and complexity of contemporary life. The book The Great Shift explores this transition, highlighting how modern narratives intertwine ecological realities and existential questions to redefine literature's role in shaping human understanding.

This seismic shift in the human condition has given rise to a new literary consciousness, one that transcends the limitations of stereotypical storytelling and ventures into the terrain of environmental collapse, planetary ethics, and collective survival. The Great Shift is a chronicle of this transformation—a literary and philosophical journey that explores how fiction is responding to the defining crisis of our time: climate change.

In a world increasingly characterized by wildfires, floods, melting glaciers, climate-induced migration, and *ecological grief*, literature has assumed a new role—not just as a mirror to reality but as a tool of resistance, resilience, and reimagination. Traditional narratives rooted in individualism are making way for stories that foreground interconnectedness between humans, non-human life, and the planet itself.

Among the most compelling genres to emerge in this era of transformation is Climate Fiction, or (Cli-Fi). Born from the intersection of speculative storytelling and environmental advocacy, (Cli-Fi) has become a powerful vehicle for exploring the profound impacts of climate change on human societies, ecosystems, and planetary futures. The genre's origins trace back to early speculative fiction, but its rise corresponds with the growing awareness of the global climate crisis. Through speculative, dystopian, and sometimes utopian lenses, (Cli-Fi) captures the fragility of the human-environment relationship and the possibilities for resilience in the face of environmental collapse.

((Cli-Fi)) is uniquely versatile, blending speculative fiction with realism, historical narratives, and even eco-fiction. It portrays scenarios ranging from post-apocalyptic survival tales to near-future realism, where communities grapple with extreme weather, resource scarcity, or climate-induced migration. Through vivid imagery, thought-provoking dialogue, and innovative storytelling, (Cli-Fi) transforms abstract environmental discourse into tangible experiences, making ecological issues resonate deeply with readers.

Beyond its literary merits, (Cli-Fi) plays a critical role in environmental advocacy and awareness. By imagining futures shaped by climate change, the genre fosters empathy, engagement, and a deeper understanding of the stakes involved. Whether depicting dystopian societies, ecological collapse, or radical adaptation, (Cli-Fi) challenges readers to confront both the consequences of inaction and the possibilities of resilience, activism, and transformation.

Several trailblazing authors have defined the genre and elevated its impact on both literature and environmental thought. Margaret Atwood's *Madd Addam Trilogy* masterfully combines speculative storytelling with ecological themes, creating narratives that are as unsettling as they are profound. Kim Stanley Robinson's *Mars Trilogy* and *New York 2140* explore scenarios of terraforming and climate adaptation, imagining futures where human ingenuity and environmental stewardship collide. *Octavia Butler's Parable of the Sower* and *Parable of the Talents* offer poignant explorations of climate-induced societal collapse and human resilience. These seminal works, among others, exemplify how (Cli-Fi) fosters critical reflection on humanity's relationship with the natural world.

As The Great Shift delves deeper into the rise of (Cli-Fi), it also introduces key concepts and frameworks essential to understanding our interconnected challenges. From "*Polycrisis*" and "*Marsification*" to "*Ecological Grief*" and "*Deep Adaptation*," these terminologies anchor readers in the urgency and scope of climate realities. Terms such as "*Climate Refugee*," "Climate Engineering," and "Climate Justice" highlight the human dimensions of the crisis, while "*Eutierria*" and "*Collapsology*" reflect philosophical and existential approaches to our collective future. The book blends literary analysis with contemporary discourse to offer readers a comprehensive understanding of both (Cli-Fi) and the broader narratives shaping humanity's response to ecological threats.

By intertwining literary creativity with ecological awareness, *The Great Shift* invites readers to embark on a transformative journey—a reimagining of storytelling that not only reflects humanity's challenges but also inspires hope and action for a sustainable future.

CHAPTER 1

Navigating the Age of *Polycrisis*: Complexity, Consciousness, and Global Literature

In an age defined by an alarming convergence of crises—from global pandemics and geopolitical conflicts to climate disasters and economic instability—the term *polycrisis* has emerged as a powerful conceptual lens to understand the intricate web of challenges confronting humanity. A *polycrisis* is not merely a cluster of concurrent crises but a condition in which these crises are causally interconnected, producing compounded and often unpredictable effects that are more devastating than the sum of their parts.

Unlike the simplistic metaphor of a "perfect storm," which suggests a temporary and unfortunate overlap of independent events, the notion of a *polycrisis* underscores the structural and systemic entanglements between multiple global domains—health, economy, environment, governance, and social cohesion. This interconnectedness means that a crisis in one area can trigger or intensify crises in others, creating a cascading chain of disruptions with long-term implications

The term *polycrisis* describes a condition in which multiple, interconnected crises occur simultaneously, with each crisis intensifying the impact of the others. Unlike isolated events, a *polycrisis* unfolds as a cascading web of disruptions that cut across economic, social, environmental, health, and political systems, amplifying systemic risks and complicating traditional problem-solving methods.

Coined by French philosopher and sociologist Edgar Morin in his 1993 work Terre-Patrie (later translated as *Homeland Earth: A Manifesto for the New Millennium*), the term *polycrisis* reflects a systemic understanding of how multiple crises interweave across domains such as economy, ecology, society, and politics. Morin argued that modern challenges are not isolated events but part of a complex web of interdependencies, where the attempt to resolve one crisis without addressing its links to others often leads to unintended consequences. His perspective was deeply rooted in complexity science and systems theory, emphasizing that the compound and interactive nature of these crises defies traditional, siloed approaches. As global challenges intensified at the turn of the 21st century—with the rise of economic globalization, the erosion of political stability, and worsening ecological degradation—Morin's framework gained renewed relevance. It has since been adopted and expanded upon by economists like Adam Tooze, policymakers such as Jean-Claude Juncker, and institutions including the World Economic Forum to articulate the unprecedented interconnectedness of contemporary global risks.

In the realm of world literature, the concept of *polycrisis* becomes a powerful lens through which authors explore the intricate relationships between these global challenges, human experience, and the pursuit of social justice. Literature, with its ability to transcend geographic, temporal, and cultural boundaries, brings the fragmented realities of the *polycrisis* into sharper focus. Through its stories, it examines how individuals, communities, and nations navigate the complexities of a fractured world, offering both a mirror to current realities and a speculative window into potential futures.

This article delves into the intersection of *polycrisis* and world literature, illuminating how literary works grapple with themes of interconnectedness, resilience, and resistance. It investigates how novels, poetry, and other narrative forms articulate the lived experiences of those most affected by these crises while imagining alternative pathways toward equity and sustainability. By doing so, world literature not only bears witness to the fractures in our systems but also prompts critical reflections on what it means to be human in an era of profound upheaval.

1. Edgar Morin's Homeland Earth: A Foundational Text for Understanding *Polycrisis*

Edgar Morin's *Homeland Earth: A Manifesto for the New Millennium* is a seminal work that provides profound insights into the concept of *polycrisis* and its implications for humanity. Co-authored with Anne Brigitte Kern, the text is a call to embrace complexity, interdependence, and systemic thinking in addressing the interconnected challenges of our time. Below is an elaborate review of the book, highlighting its foundational ideas and relevance.

Morin introduces the term *polycrisis* to describe the intricate web of global challenges that cannot be understood or resolved in isolation. He emphasizes that crises such as climate change, economic instability, social inequality, and geopolitical tensions are deeply interconnected, forming a "complex intersolidarity of problems". This interconnectedness demands a shift from reductionist thinking to a systemic approach that acknowledges uncertainty and ambiguity as sources of creativity and transformation.

The book advocates for a paradigm of complex thought, which Morin describes as essential for navigating the Anthropocene—a geologic epoch defined by human impact on the planet. By integrating ecological, social, and cultural dimensions, Morin's framework challenges the fragmented perspectives that often dominate policy and decision-making.

One of the central themes of *Homeland Earth* is the idea that humanity must recognize itself as a planetary and biospheric entity. Morin argues that the sharp distinctions between humans and the rest of the biosphere, inherited from Renaissance humanism and Cartesian philosophy, run counter to the interconnected nature of existence. He calls for a "cosmic polyidentity" that weaves connections between humans and the myriad links that constitute the chain of life.

This perspective urges individuals and societies to re-anchor their consciousness in the workings of an interdependent planet. Morin's vision is not merely ecological but deeply philosophical, advocating for a richer sense of belonging that transcends anthropocentric delusions of grandeur.

Morin's paradigm of systemic complexity is particularly relevant to discussions of sustainability. He critiques the dichotomy between economics and the environment, proposing instead a network of solidarity and cooperation among diverse actors, institutions, and communities. This approach aligns with his broader vision of eco-self-organization, where humanity learns to live in harmony with the chaotic forces of nature rather than attempting to dominate them.

The book also explores the role of cultural and epistemological biases in shaping environmental policies. By integrating concepts of eco-organization and environmental complexity, Morin provides a blueprint for rethinking sustainability in contemporary times.

Morin's work is not just a manifesto for ecological action but also a profound philosophical inquiry into the nature of human identity and responsibility. He challenges readers to "know thyself" by exploring humanity's relationship with the ecological forces that sustain life. This call to reintegrate ourselves into the larger biotic community is both a critique of modern alienation and a vision for a more interconnected future.

The text also addresses the moral imperative to act in solidarity with those most affected by global crises. Morin's emphasis on complexity and interdependence underscores the need for ethical frameworks that prioritize reciprocity and mutual care.

Homeland Earth is a foundational text for understanding *polycrisis* and its implications for humanity. Morin's call for complex thought, planetary consciousness, and systemic cooperation offers a transformative vision for navigating the challenges of the Anthropocene. By weaving together ecological, social, and philosophical dimensions, the book invites readers to rethink their place in the world and their role in shaping its future.

This work is not just a critique of existing paradigms but a hopeful blueprint for reimagining human identity and action in an era of profound interconnectedness. It reminds us that the solutions to global challenges lie not in isolated efforts but in embracing the complexity and interdependence that define our shared existence.

2. Adam Tooze and the Anatomy of *Polycrisis*: A Review of Shutdown: How Covid Shook the World's Economy

Adam Tooze's *Shutdown: How Covid Shook the World's Economy* is a compelling and incisive analysis of the global economic upheaval triggered by the COVID-19 pandemic. Through his characteristic clarity and rigor, Tooze dissects the interconnected crises that emerged during the pandemic, offering readers a nuanced understanding of the *polycrisis* that continues to shape our world. Below is an elaborate review of the book:

Tooze begins by chronicling the unprecedented economic collapse that followed the outbreak of COVID-19. From the initial shockwaves in financial markets to the cascading effects on industries such as tourism, oil, and hospitality, the book captures the sheer scale of disruption. Tooze highlights how governments worldwide responded with extraordinary measures, including massive fiscal stimulus packages and central bank interventions, to prevent total economic collapse.

What sets Shutdown apart is its ability to connect the dots between seemingly disparate events. Tooze frames the pandemic as a *polycrisis*—a convergence of health, economic, and social crises that exposed the vulnerabilities of global systems. This interconnectedness is central to his analysis, emphasizing that the pandemic was not just a public health emergency but a systemic shock that reshaped the global economy.

Tooze's exploration of *polycrisis* is one of the book's most insightful contributions. He argues that the pandemic revealed the fragility of interconnected systems, where disruptions in one area, such as supply chains, rippled across economies and societies. The book examines how pre-existing inequalities were exacerbated, with vulnerable populations bearing the brunt of the crisis.

Tooze also critiques the global response to the pandemic, questioning whether the measures taken were sufficient to address the systemic nature of the crisis. He highlights the uneven recovery across nations, pointing to the disparities between wealthy countries with access to vaccines and resources and poorer nations struggling to cope.

A significant portion of the book is dedicated to analyzing the role of financial markets and institutions during the pandemic. Tooze delves into the unprecedented interventions by central banks, which injected liquidity into markets to stabilize economies. He examines the implications of these actions, including the long-term consequences for debt, inflation, and economic inequality.

Tooze's analysis is both technical and accessible, making complex economic concepts understandable for a broad audience. He provides a critical perspective on the role of financial institutions, questioning their ability to address systemic risks in an era of *polycrisis*.

While Shutdown is primarily an account of the pandemic's economic impact, it also offers a vision for the future. Tooze calls for a rethinking of global systems to address the interconnected challenges of the Anthropocene. He advocates for greater international cooperation, systemic resilience, and a shift away from short-term thinking.

The book underscores the importance of addressing inequality and sustainability as central pillars of recovery. Tooze's vision is both pragmatic and hopeful, emphasizing the need for collective action to navigate the complexities of a *polycrisis* world.

Adam Tooze's Shutdown is a masterful exploration of the economic and systemic challenges posed by the COVID-19 pandemic. By framing the crisis as a *polycrisis*, Tooze provides readers with a deeper understanding of the interconnected nature of global challenges. His analysis is both rigorous and accessible, making the book an essential read for anyone seeking to understand the economic and social implications of the pandemic.

The book is not just a historical account but a call to action, urging policymakers, institutions, and individuals to rethink their approaches to resilience and sustainability. Tooze's insights are particularly relevant as the world continues to grapple with the ripple effects of the pandemic and other emerging crises.

3. *Flight Behavior by* Barbara Kingsolver.

Barbara Kingsolver's *Flight Behavior* (2012) is a multifaceted novel that explores the complex interplay between human lives and the natural world, using the phenomenon of a disrupted monarch butterfly migration as a powerful metaphor for climate change and personal transformation.

Set against the lush yet rugged backdrop of Appalachia, the novel introduces us to Dellarobia Turnbow—a young, disillusioned farm wife whose life has been circumscribed by early marriage, the monotony of rural existence, and the weight of unfulfilled dreams. Initially confined by the limited expectations of her community and her circumstances, Dellarobia's encounter with an awe-inspiring natural spectacle provides the catalyst for a profound internal awakening. Kingsolver's vivid descriptions not only paint a portrait of a unique geographical and cultural setting but also invite readers to consider how the local environment reflects larger global shifts.

At the heart of *Flight Behavior* is an extraordinary event: a massive congregation of monarch butterflies appearing in a place and season where such a display should be impossible. The butterflies, whose traditional migratory pattern has been thrown into disarray by climate change, serve as both a literal and symbolic manifestation of the ecological disruptions that define the modern world. For Dellarobia, this unexpected invasion is much more than a natural anomaly—it is an invitation to reexamine her life's limitations and the societal forces that have kept her tethered. As scientific curiosity converges with local folklore, Kingsolver weaves together strands of personal narrative and environmental observation, building a story that is as much about inner transformation as it is about the unsettling effects of global warming.

One of the novel's striking features is how it elevates environmental change to the level of personal and communal metamorphosis. The monarch butterflies symbolize resilience amid chaos, beauty intertwined with impermanence, and nature's ability to disrupt even the most rigid human structures. Through the disrupted migration, Kingsolver communicates a broader message about the interconnectedness of life: local environmental changes mirror and are inextricable from the global shifts in climate. Alongside this environmental allegory, the book examines themes of female empowerment, self-discovery, and the tensions between tradition and modern scientific understanding. The clash between entrenched beliefs and emerging ecological realities creates a narrative space where personal redemption and environmental activism intersect, urging readers to consider how local actions are part of a larger, global story.

Dellarobia stands out as a richly rendered character whose inner journey is as compelling as the external ecological drama unfolding around her. Kingsolver portrays her with empathy and nuance, capturing the struggle of a woman caught between the expectations of her small, tight-knit community and the stirrings of a deeper, more critical awareness about the modern world. Interactions with other figures—be they skeptical family members, the well-intentioned yet sometimes patronizing community, or the scientists who come to study the phenomenon—highlight the conflicting responses to change. The novel challenges its characters (and in turn, its readers) to balance romantic lore and pragmatic science, personal belief, and inconvenient truth. This dynamic not only enriches the narrative but also offers a pointed cultural commentary on how societies confront inconvenient environmental realities.

Kingsolver's prose is both lyrical and incisive. Her careful attention to detail in describing the Appalachian landscape and the ephemeral beauty of the monarch butterflies draws readers into a world where every natural occurrence resonates with deeper significance. The narrative is layered with symbolism, inviting multiple interpretations—whether one views the butterfly invasion as a harbinger of ecological crisis or as a moment of personal redemption. Beyond its narrative scope, *Flight Behavior* has sparked broader conversations about climate change, environmental stewardship, and the ways literature can capture the urgency of the Anthropocene. Critics have lauded the novel for its balanced approach to complex issues, making it not only a rich literary work but also an instrument for ecological awareness and social reflection.

Ultimately, *Flight Behavior* is more than just a story about a natural phenomenon; it is a meditation on the transformative power of nature and the resilience of the human spirit. By intersecting the personal journey of its protagonist with the global challenge of climate change, the novel encourages us to reconsider how individual lives are interwoven with broader ecological processes. Kingsolver invites readers to see that the seemingly disparate threads of personal dissatisfaction, local tradition, and global environmental crisis are part of a larger tapestry that demands reflection, engagement, and, ultimately, action.

4. *The Ministry of Utmost Happiness by* Arundhati Roy.

Arundhati Roy's *The Ministry of Utmost Happiness* (2017) is an expansive, audacious narrative that defies easy categorization. It interweaves multiple lives, geographies, and temporalities to offer a panoramic view of modern India—a landscape marred by historical wounds, social exclusions, and political tumult. The novel is both a call to recognize the marginalized and a poetic lament for a nation grappling with its contradictions.

Roy's narrative unfolds in a non-linear mosaic that spans vivid episodic portraits and sprawling historical vignettes. The story moves fluidly between the lives of disparate characters—from the resilient transgender protagonist, Anjum, whose life embodies unattainable dreams and fierce defiance, to figures embroiled in political and regional strife. This multi-stranded approach mirrors the chaotic pulse of modern India, allowing Roy to create a work whose structure is as much a statement on fragmented identity as it is a literary technique.

At its core, the novel is an exploration of identity in its myriad forms. Roy scrutinizes the fissures in Indian society—racial, gender, religious, and class—that force individuals to inhabit lives on the fringe. Her characters grapple with the perennial search for belonging in a country where history and destiny are tightly wound. For many, such as Anjum, life is a relentless quest for recognition and dignity in spaces that systematically ignore them. Roy doesn't offer easy resolutions; instead, she lays bare the painful process of self-definition amid adversity, crafting a narrative that is as empathetic as it is critical.

The *Ministry of Utmost Happiness* is as much a work of political protest as it is a novel. Roy's keen observations on state power, imperial legacies, and the exploitation of the disenfranchised infuse the text with an urgency that challenges conventional narratives of national identity. The novel spans from intimate personal sorrows to overtly political expressions, such as the struggles against state-sanctioned violence, the legacy of partition, and the exploitation endemic to modern governance. These elements underscore her argument that the personal and the political are inextricably intertwined, provoking readers to re-examine their assumptions about justice and history.

Roy's prose is both lyrical and unflinching. Her language is rich with metaphor and abrasion, evoking the spiritual, the sensual, and the sorrowfully tangible landscapes of India. The ministry in the title itself becomes an emblem—a repository of aspirations and despair, where states and personal lives coalesce in a relentless search for meaning. The symbolism of spaces, from the crowded margins of society to the alleys of a postcolonial nation, serves as a powerful counterpoint to the often oppressive political realities she describes. Each narrative layer, whether a fleeting relationship or a monumental event, resonates with a scaled metaphor, reflecting the multifaceted nature of human experience.

One of the most compelling aspects of Roy's work is its deep commitment to the voices that are typically rendered invisible. By centering characters such as Anjum—a figure who embodies the struggle of gender nonconformity and societal rejection—Roy challenges the normative narratives imposed by a dominant culture. Though scarred by exclusion and violence, her characters display a tenacity that transforms their pain into a kind of fierce beauty. The novel thus becomes an ode to the resilience found in life's margins, urging readers to reconsider who is deemed significant in the national story.

In summation, The Ministry of Utmost Happiness is a daring exploration of modernity's fissures. Its elaborate narrative structure, thematic depth, and unapologetic political critique invite readers to confront uncomfortable truths about identity, belonging, and state power. Roy's work is not merely a story but a provocation that asks: Can a nation rediscover its humanity when so many of its people remain unheard? How do we reconcile the personal with the political in our collective search for justice?

This intricate interplay of personal anguish and political activism resonates profoundly in today's world, where the margins continue to speak—and where literature remains a potent medium for social transformation.

Towards a New Consciousness in the Age of Polycrisis

To confront the age of *polycrisis*, we must move beyond narrow disciplinary silos and cultivate an integrative imagination—one capable of holding paradoxes, connecting dots, and seeing the whole in the parts. We need new ways of knowing, new modes of collaboration, and new ethical frameworks that prioritize resilience, adaptability, and justice. The choice before us is stark: either continue down the path of fragmentation and reaction or rise to the challenge of complexity with courage, creativity, and collective intelligence.

In this moment of converging crises, the works of thinkers like Morin and Tooze are not merely academic—they are maps for survival, for solidarity, and for reimagining the world anew.

CHAPTER 2

Sick Planet, Shattered Stories: The *Triple Planetary Crisis* in Climate Fiction.

In recent decades, the world has been confronted with a compounding environmental emergency described as the *triple planetary crisis*, a term adopted by the United Nations to refer to the intersecting threats of climate change, pollution, and biodiversity loss. These issues, while distinct in origin, are deeply interwoven and collectively threaten the integrity of ecosystems, the health of populations, and the sustainability of global economies (UNEP). The urgency of this crisis is not merely ecological but existential, demanding both systemic reform and individual commitment. As science-based policy struggles to keep pace with environmental deterioration, literature, particularly climate fiction, has emerged as a vital medium for raising awareness and fostering ecological consciousness.

The *triple planetary crisis* comprises three interrelated environmental threats:

1. Climate Change: A consequence of greenhouse gas emissions primarily from fossil fuel use, climate change manifests in rising temperatures, extreme weather events,

sea level rise, and ecosystem disruption. It is widely acknowledged as the most urgent environmental threat of our time (UNEP).

2. Pollution: From plastic waste and chemical contamination to air and water degradation, pollution is a ubiquitous crisis that affects all forms of life. It not only harms wildlife but also poses severe health risks to humans, particularly vulnerable communities (UNEP).

3. Biodiversity Loss: Driven by habitat destruction, overexploitation, and invasive species, biodiversity loss reduces the resilience of ecosystems and undermines essential services such as pollination, water purification, and climate regulation.

These crises are interconnected. Climate change accelerates biodiversity loss, while pollution exacerbates ecological fragility. Together, they represent a systemic failure of anthropogenic activity to align with planetary boundaries (Steffen et al.).

In response, modern literature, particularly speculative genres like climate fiction ((Cli-Fi)), has increasingly embraced these urgent themes, presenting them through imaginative storytelling that resonates on both emotional and intellectual levels.

Fiction can transcend cultural and linguistic boundaries, reaching audiences that scientific texts may not. By embedding climate realities into stories, literature raises consciousness about environmental crises in accessible ways. Stories evoke a range of powerful emotions—grief, empathy, urgency, and hope—that engage readers on a visceral level. These emotional responses often lead to a deeper understanding of ecological loss. Many readers, especially among the youth, report feeling inspired to change behaviours, support environmental advocacy, or engage in policy debates after reading climate-centered literature.

Literature challenges human-centered views and cultivates ecological awareness, prompting readers to recognize their interconnectedness with the planet. As environmental challenges vary across regions, narratives from diverse cultural contexts promote empathy and shared understanding. This global literary exchange fosters a sense of planetary solidarity.

Contemporary works of fiction often place this crisis at the core of their narratives, using it to shape plotlines, drive character arcs, and infuse stories with a sense of ethical immediacy. These literary pieces frequently translate complex scientific issues into personal, emotionally compelling narratives, criticizing dominant social structures, such as capitalism and unchecked technological advancement, that contribute to ecological damage. The also envision alternate futures, offering either cautionary tales or hopeful blueprints for what might come next.

Authors such as Margaret Atwood, Peter Heller, Emily St. John Mandel, and Kim Stanley Robinson use speculative narratives to explore the implications of environmental collapse, envisioning dystopian and resilient futures shaped by the *triple planetary crisis*.

This chapter examines how four major works of climate fiction—*The Year of the Flood* (2009), *The Dog Stars* (2012), *Station Eleven* (2014), and *The Ministry for the Future* (2020)—engage with the *triple planetary crisis*. These texts not only illustrate the consequences of ecological neglect but also function as imaginative blueprints for survival, resistance, and hope in the face of planetary transformation.

1. The Year of the Flood by Margaret Atwood.

Set in a dystopian future, The Year of the Flood—the second novel in Atwood's Madd Addam Trilogy—unfolds in a world ravaged by ecological devastation and dominated by corporate power. The plot centres on two women, Toby and Ren, survivors of a man-made pandemic and followers of an eco-religious group called God's Gardeners. The story oscillates between individual survival and the collapse of natural and societal systems.

In *The Year of the Flood*, Atwood constructs a dystopian future where unchecked corporate greed, genetic engineering, and environmental degradation culminate in ecological and societal collapse. The novel presents a world ravaged by pollution and climate instability, where biodiversity has been replaced by synthetic lifeforms engineered for profit. The narrative follows members of God's Gardeners, an eco-spiritual sect that resists consumerism and promotes sustainability. Atwood critiques the commodification of nature and warns against the ethical and ecological dangers of biotechnology. Her depiction of a world poisoned by pollution and climate manipulation aligns directly with the *triple planetary crisis*, offering a stark vision of what may lie ahead if ecological balance is not restored (Atwood).

The narrative foregrounds the consequences of climate instability, including rising sea levels, erratic weather, and resource scarcity. Urban environments become increasingly uninhabitable due to environmental stressors. Food and water scarcity threaten survival, especially for the marginalized. The God's Gardeners promote carbon-conscious living and critique the dangers of fossil fuel dependence. Atwood mirrors real-world concerns, highlighting how climate disruption exacerbates inequality and disproportionately harms vulnerable populations.

Margaret Atwood's *The Year of the Flood* serves as a powerful commentary on the *triple planetary crisis*, weaving together themes of climate change, pollution, and biodiversity loss. Her narrative critiques the systems responsible for ecological collapse while also presenting possibilities for ethical resistance and regeneration. Such literature does not merely imagine dystopian outcomes—it invites readers to reflect, respond, and rethink their relationship with the Earth. As stories like these continue to gain prominence, they play an essential role in shaping ecological consciousness and motivating change.

2. *The Dog Stars by* Peter Heller.

Peter Heller's *The Dog Stars* presents a compelling vision of a world reshaped by ecological collapse and a global pandemic. Set in a post-apocalyptic landscape, the novel follows Hig, a lone survivor, as he navigates the aftermath of environmental ruin and societal disintegration. Through his journey, the novel subtly critiques humanity's impact on nature and the fragility of human connections in a broken world.

The narrative subtly weaves in elements of the *triple planetary crisis*—climate change, pollution, and biodiversity loss—portraying them not as isolated phenomena but as deeply linked forces driving collapse. Although not overtly discussed, the novel suggests that shifts in the climate may have contributed to the outbreak of disease, emphasizing the interdependence of human health and environmental stability. The degradation of natural resources, especially the scarcity of clean water and the collapse of ecosystems, illustrates the severe consequences of pollution and unsustainable practices. The haunting silence in nature, such as the lack of birdsong, becomes a powerful symbol of diminished biodiversity, highlighting how the disappearance of species disrupts ecological balance.

Heller's work also delves into the psychological toll of environmental destruction. Characters experience a profound sense of loss—not just of people, but of the world they once knew. This *ecological grief* reflects a deeper disconnection from nature, suggesting that human identity and well-being are intrinsically linked to the health of the environment. Heller evokes a melancholic nostalgia for the pre-crisis world, revealing how intimately human identity is tied to ecological health (Heller).

Peter Heller's *The Dog Stars* artfully depicts the ways in which environmental breakdown, disease, and emotional disconnection intersect. Through its post-apocalyptic lens, the novel not only critiques unsustainable human behaviour but also inspires reflection, action, and solidarity. As a piece of climate fiction, it calls attention to the urgent need for ecological awareness and collective resilience in the face of planetary threats.

3. *Station Eleven by Emi*ly St. John Mandel

Emily St. John Mandel's *Station Eleven* offers a profound meditation on the *triple planetary crisis*—climate change, pollution, and the loss of biodiversity. Though the novel's central narrative unfolds in the wake of a devastating pandemic, it quietly integrates environmental themes, crafting a layered vision of a civilization that has unravelled. Mandel's depiction of a post-collapse world subtly critiques the ecological damage wrought by human progress.

In the aftermath of societal breakdown, *Station Eleven* presents a world stripped of its technological scaffolding and modern-day luxuries. This enforced return to a simpler existence allows the remaining population to re-engage with the natural environment. The absence of industrial pollution, a direct consequence of the pandemic's societal reset, restores a kind of ecological balance. Mandel uses this scenario to critique the ecological toll of human advancement, subtly emphasizing the value of living in harmony with nature and the urgent need for responsible environmental practices.

Mandel's narrative underscores how modern civilization's relentless pursuit of advancement has exacted a severe ecological cost. The novel suggests that material progress has often gone hand in hand with environmental exploitation—manifesting in climate disruptions, species extinction, and rampant pollution. Through its interwoven themes of cultural survival and environmental decay, *Station Eleven* emphasizes that sustainability is not merely about resource conservation but also about nurturing human creativity and fostering meaningful social bonds.

The novel places considerable emphasis on the role of memory and cultural continuity in grappling with environmental degradation. Characters frequently reflect on the world that existed before the collapse, drawing attention to what has been lost—not only in terms of infrastructure but also in cultural and environmental richness. These recollections function as a call to safeguard not just the planet but the collective human experience that accompanies it. Mandel thereby proposes a holistic vision of sustainability—one that preserves both the natural world and the stories, traditions, and values of its inhabitants.

Station Eleven has struck a chord with audiences across the globe, particularly among those invested in ecological and climate issues. Its subtle yet compelling portrayal of environmental themes has contributed to growing discourse around sustainability and ecological responsibility. By framing the planetary crisis through the lens of fiction, Mandel encourages a deeper emotional and intellectual engagement with environmental challenges, prompting readers to become more conscientious global citizens.

Ultimately, Mandel's novel serves as a literary beacon, shedding light on the grave realities of ecological decline while offering a vision for change. It challenges readers to examine their environmental footprint and to reimagine a future rooted in sustainability and care for the Earth. As a piece of climate fiction, *Station Eleven* functions not only as a narrative of survival but also as a catalyst for ecological awareness and action, urging readers to contribute to a more resilient and environmentally just future.

From Crisis to Consciousness: The Literary Reckoning with Planetary Peril

The *triple planetary crisis* represents a convergence of environmental catastrophes that challenge the very foundations of human and non-human life on Earth. Climate change, pollution, and biodiversity loss are not isolated phenomena; they are mutually reinforcing threats born of unsustainable human behaviour and systemic negligence. Through the lens of climate fiction, authors like Atwood, Heller, Mandel, and Robinson render these abstract global crises tangible, emotional, and urgent. Their narratives serve as both cautionary tales and imaginative roadmaps, urging readers to reckon with the consequences of ecological collapse and to envision alternative futures grounded in sustainability and resilience. As literature bridges the gap between science and empathy, it becomes an essential tool in confronting the *triple planetary crisis*, prompting reflection, resistance, and collective action.

CHAPTER 3

Marsification: Reimagining Humanity's Future Beyond Earth

Throughout history, humanity has been driven by an insatiable curiosity and an urge to explore uncharted territories. From the early days of maritime voyages to the space age's monumental strides, humans have consistently sought to expand their horizons. Today, as Earth faces escalating environmental and social pressures, this quest for new possibilities has led us to contemplate life beyond our planet. At the forefront of this visionary endeavor is *Marsification*—the theoretical transformation of Mars into a hospitable environment that can sustain human life. This concept not only challenges our technological ingenuity but also raises profound ethical and philosophical questions about our role in the cosmos. By blending scientific ambition with speculative imagination, *Marsification* inspires hope for humanity's resilience and adaptability, while also prompting critical reflection on the future of our planet.

The concept of *Marsification*—the theoretical transformation of Mars into a habitable environment for humans and Earth-like species—has captivated the imagination of scientists, environmentalists, and science fiction authors alike. Rooted in both scientific ambition and literary exploration, *Marsification* has emerged as a powerful symbol of hope, survival, and renewal in the face of Earth's environmental and socio-political uncertainties. As global challenges such as climate change, overpopulation, and dwindling natural resources intensify, the idea of turning our neighbouring planet into a second home no longer feels like a mere fantasy, but rather a scientific frontier and ethical quandary worth investigating.

The term *Marsification* originates from the name of the Red Planet—Mars—which has long been a focal point for scientific curiosity, philosophical speculation, and speculative fiction. The concept is closely tied to the broader field of terraforming, a term first coined by science fiction author Jack Williamson in a 1942 short story, "Collision Orbit," published *in Astounding Science Fiction* (Williamson, 1942). Terraforming refers to the process of deliberately altering the atmosphere, temperature, surface topography, or ecology of a planet to resemble that of Earth. Within this framework, *Marsification* represents a Mars-specific adaptation, encompassing both the technological means to transform the planet and the socio-political aspirations driving such efforts.

Several visionaries and thought leaders have been instrumental in developing and popularizing the idea of *Marsification*, contributing to both its scientific and cultural narrative.

1. Robert Zubrin: An American aerospace engineer, Zubrin is the founder of the Mars Society and author of The Case for Mars (1996). He has argued that human settlement on Mars is not only feasible but imperative for the long-term survival of civilization. His Mars Direct proposal outlines a cost-effective strategy for manned missions and colonization using in-situ resource utilization (Zubrin, 1996).

2. Elon Musk: The founder and CEO of SpaceX, Musk has significantly advanced the conversation around interplanetary colonization. Through the development of the Starship program, Musk envisions a future where Mars hosts self-sustaining human settlements as part of a multi-planetary survival strategy (Musk, 2017). His public statements and technological initiatives have brought *Marsification* into mainstream discourse, influencing public policy and private investment.

Marsification carries profound implications that span scientific, environmental, technological, and sociocultural domains. As a frontier of space exploration, it pushes the boundaries of what is technologically possible, offering valuable insights into planetary science, astrobiology, and the extremes of human endurance. Simultaneously, it emerges as a symbolic and practical response to Earth's ecological decline—a contingency plan that envisions a future beyond our deteriorating biosphere. The pursuit of Mars colonization fuels technological innovation, particularly in developing life-support systems, propulsion technologies, and radiation shielding, many of which have direct applications in managing extreme or disaster-prone conditions on Earth. Beyond the scientific and practical, *Marsification* holds deep psychological and ethical resonance. It acts as a mirror to human ambition and adaptability, compelling us to reflect on the values we carry and the kind of societies we aim to recreate—or consciously leave behind—in a new world.

Despite its futuristic appeal, *Marsification* is fraught with complex technical, ethical, and economic challenges. Technically, the Martian environment poses formidable obstacles—its low atmospheric pressure, extreme cold, and lack of breathable air demand advanced engineering solutions. Technologies for creating artificial biospheres, atmospheric processing, and sustaining long-term human habitation remain largely experimental (NASA, 2022). Ethically, the prospect of altering another planet raises profound concerns. Should humanity intervene in an alien ecosystem that might harbor microbial life? If so, what responsibilities do we bear in preserving any potential Martian biosphere in its natural state? Economically, the immense costs associated with terraforming or colonizing Mars risk diverting attention and resources from urgent issues on Earth, such as poverty, healthcare, and climate change. Critics caution that *Marsification* may function more as an escapist fantasy than a viable solution to the ecological crises we face at home.

The Literary Legacy of Marsification

Long before Mars became a practical target for exploration, writers used it as a metaphor for imperialism, isolation, and the unknown. Four major literary figures have especially shaped the evolving narrative of *Marsification.* The speculative genre of climate fiction, or (Cli-Fi), has become a powerful vehicle for exploring the implications of *Marsification.* In these narratives, the Red Planet becomes both a setting and a symbol, often representing a last refuge from ecological collapse on Earth. Notably, these stories critique both the promise and perils of colonization and technological hubris.

The concept of "*Marsification*"—the transformation of Mars into a habitat suitable for human life—has been a compelling theme in science fiction literature. Four seminal works that explore this idea are H.G. Wells's *The War of the Worlds,* Ray Bradbury's *The Martian Chronicles*, and Kim Stanley Robinson's *Red Mars, and Geoffrey A. Landis's Mars Crossing.* Each of these narratives offers a distinct perspective on *Marsification*, exploring themes of colonization, environmental transformation, and the human condition.

1.H.G. *Wells's The War of the Worlds.*

Published in 1898, H.G. *Wells's The War of the Worlds* presents an early exploration of Martian interaction with Earth, albeit through the lens of invasion rather than human colonization of Mars. The novel depicts Martians as technologically superior beings who invade Earth, leading to the collapse of human society's social order. This narrative serves as a critique of British imperialism, drawing parallels between the Martian conquest of Earth and European colonization of other lands. Wells challenges the notion of human supremacy by positioning humans as the subjugated species, prompting readers to reconsider humanity's place in the universe and its treatment of other beings.

The Martians, armed with superior technology, ravage human civilization—offering a stark allegory for European imperialism. In this reversed colonial encounter, *Marsification* becomes a metaphor for environmental and cultural domination. The Martians' downfall, due to Earth's bacteria, serves as a cautionary tale about underestimating the intricacies of alien ecosystems.

While *The War of the Worlds* does not depict *Marsification* in the traditional sense, it inversely examines the consequences of a more advanced civilization imposing its will upon another. The Martians' attempt to transform Earth into a habitat suitable for themselves mirrors the concept of humans altering Mars for their own habitation. This inversion serves as a cautionary tale about the ethical implications of colonization and environmental manipulation.

2. Ray Bradbury's The Martian Chronicles.

Ray Bradbury's T*he Martian Chronicles*, published in 1950, offers a series of interconnected stories that chronicle humanity's colonization of Mars. Bradbury uses *Marsification* as a metaphor to critique human tendencies toward environmental exploitation and cultural imperialism. In the narrative, humans arrive on Mars and begin to reshape its environment to resemble Earth, often disregarding the existing Martian culture and ecosystem. This reflects historical instances of colonization where indigenous cultures were suppressed or obliterated by colonizers.

Bradbury offers a poetic and philosophical exploration of *Marsification*. His stories portray Earthlings colonizing Mars, only to replicate the violence, racism, and ecological destruction of their home planet. The Martian Chronicles questions whether humanity can evolve beyond its destructive tendencies or if it is doomed to repeat them on alien soil.

One poignant example is the story "-And the Moon Be Still as Bright," where the character Spender laments the destruction of Martian civilization due to human carelessness and disease. This narrative underscores the moral and ethical dilemmas associated with *Marsification*, emphasizing the need for respect and preservation of alien cultures and environments. Bradbury's work serves as a cautionary tale, urging readers to reflect on the consequences of imposing human will upon foreign worlds.

3.Kim Stanley Robinson's ***Red Mars*****.**

Published in 1992, Kim Stanley Robinson's *Red Mars* provides an in-depth and scientifically grounded exploration of *Marsification*. The novel follows the first hundred colonists as they embark on the monumental task of terraforming Mars. Robinson delves into the scientific, political, and ethical challenges of transforming an alien environment to support human life.

Robinson's *Mars Trilogy* remains the most scientifically grounded and ethically rich portrayal of *Marsification*. By examining governance, ecology, and human identity on Mars, Robinson moves beyond simple survivalism or conquest, suggesting a path toward sustainable coexistence with alien landscapes.

The narrative presents a dichotomy between two factions: the "Reds," who advocate for preserving Mars in its pristine state, and the "Greens," who support terraforming to make the planet habitable for humans. This conflict mirrors real-world debates about environmental conservation versus development. Robinson's detailed portrayal of the terraforming process, from the construction of habitats to the alteration of the Martian atmosphere, highlights the immense technological and ethical considerations involved in *Marsification*.

Moreover, *Red Mars* examines the socio-political ramifications of colonizing a new world, including issues of governance, economic disparity, and cultural identity. The novel suggests that while *Marsification* offers opportunities for a fresh start, it also risks replicating Earth's societal flaws if not approached with caution and foresight.

Through these works, Wells, Bradbury, and Robinson offer nuanced explorations of *Marsification*, each highlighting different facets of humanity's potential expansion into space. Wells provides a cautionary tale about the perils of imperialism and the assumption of superiority. Bradbury critiques human tendencies toward environmental and cultural domination, urging respect for alien worlds. Robinson presents a comprehensive examination of the scientific and ethical complexities of terraforming, emphasizing the need for thoughtful consideration of the consequences. Collectively, these narratives serve as both reflections on human nature and guides for approaching the prospect of *Marsification* with humility and responsibility.

4. Geoffrey A. Landis's *Mars Crossing*.

Published in 2000, *Mars Crossing* by Geoffrey A. Landis presents a gripping tale of endurance and discovery, with the Martian environment playing a central role in the transformation of its human visitors. Landis, known for his dual expertise as a NASA scientist and science fiction writer, grounds the narrative in scientific authenticity. The novel earned the Locus Award for Best First Novel in 2001 and is widely recognized for its accurate portrayal of Martian landscapes and mission logistics (Nicholls, 2001)

The narrative centres on a group of astronauts stranded on Mars, who must traverse its unforgiving terrain in a desperate bid for survival. Landis paints a detailed and realistic picture of the Martian surface—dust-choked plains, towering volcanoes, and unpredictable weather patterns. The planet itself becomes a dynamic force in the story, shaping both the physical journey and the internal struggles of the characters.

Drawing heavily from the Mars Direct concept, Landis imagines a mission where astronauts produce return fuel from Martian resources, offering a plausible and well-researched vision of interplanetary travel. The novel exemplifies *Marsification* by illustrating how the planet influences and alters human behaviour, revealing both vulnerability and resilience in extreme conditions (Nicholls, 2001).

Although some critics argue the characters are secondary to the plot, the novel delves into the psychological challenges faced by the crew. The story addresses themes of loss, hope, and interpersonal tension, showing how the Martian environment acts as a crucible that reveals the true nature of everyone.

Themes of exploration, survival, and human adaptability are skillfully woven into the story. Landis employs a writing style that is both technically precise and emotionally resonant. The novel's pace fluctuates between intense action sequences and quieter moments of introspection, maintaining engagement while inviting deeper reflection.

Landis's *Mars Crossing* stands out in the hard science

fiction genre, offering a nuanced portrayal of how the Martian setting reshapes human identity and experience. Through its realistic science and compelling narrative, the novel offers a powerful contribution to the literary exploration of Mars.

Marsification as a Mirror to Humanity

The discourse surrounding *Marsification* is not merely a speculative indulgence or scientific ambition; it reflects our deepest fears, hopes, and questions about survival and ethics. As we stand at the crossroads of the climate crisis and space exploration, *Marsification* offers both a narrative framework and a practical inquiry into what it means to be human.

Whether seen through the lens of scientific possibility, ethical responsibility, or literary metaphor, *Marsification* forces us to confront the consequences of our actions and the future we wish to build, whether on Earth or elsewhere in the cosmos.

As we continue to explore this compelling concept in fiction and real-world science, *Marsification* challenges us to answer a simple but profound question: *If given a second chance on another planet, can humanity do better?*

CHAPTER 4

Climate Engineering and the Role of Literature in the Anthropocene

The Anthropocene, the epoch defined by humanity's transformative impact on the planet, has ushered in urgent debates on how best to respond to the accelerating climate crisis. As strategies evolve from mitigation and adaptation to direct intervention, climate engineering—or geoengineering—has emerged as a controversial proposition to alter Earth's natural systems. From reflecting sunlight using aerosols to removing carbon dioxide via advanced technologies, these approaches offer a glimpse into the potential of large-scale solutions to counteract global warming. Yet, such proposals are fraught with scientific uncertainties, ethical dilemmas, and governance challenges, raising pivotal questions about equity, justice, and humanity's responsibility toward nature.

Amid these technical and political debates, literature provides an invaluable space for exploring the ideologies, possibilities, and risks associated with climate engineering. World literature, with its ability to traverse cultural and temporal boundaries, invites readers to confront the moral and social dimensions of geoengineering while fostering critical reflection on humanity's relationship with the environment. Through its imaginative narratives, literature challenges the logic of "playing God," interrogates the socio-political consequences of climate interventions, and offers alternative visions that question dominant paradigms of power and control.

This article delves into the intersection of *climate engineering* and world literature, examining how literature shapes and contests public discourses surrounding geoengineering. By addressing themes of visual politics, ethical debates, and justice considerations, the discussion highlights how literary narratives enable a deeper engagement with the complexities of climate governance in the Anthropocene. Through a fusion of realism and imagination, literature becomes a vital tool for expanding our understanding of humanity's role in shaping the planet's future.

The Literary Response: Imagining the Engineered Earth

World literature provides a dynamic framework for reflecting on *climate engineering* not only as a scientific endeavor but as a complex socio-cultural phenomenon. Through its diverse genres, voices, and perspectives, literature interrogates the human ambition to manipulate natural systems and delves into the ethical, philosophical, and existential questions that arise from this impulse.

Writers like Kim Stanley Robinson and Margaret Atwood craft narratives that vividly illustrate the promises and perils of climate engineering. Robinson's The Ministry for the Future imagines solar radiation management (SRM) as a response to catastrophic heatwaves, probing both the urgency of intervention and its cascading consequences. Atwood's *Madd Addam* Trilogy explores engineered environments as symbols of ecological upheaval and human hubris, offering cautionary tales about unintended outcomes. These speculative works serve as mirrors and warnings, encouraging readers to critically evaluate techno-fixes within broader systems of power and inequity.

Literature from the Global South and indigenous authors enriches the discourse by challenging Western-centric approaches to geoengineering. Amitav Ghosh's The Great Derangement critiques climate narratives that marginalize non-Western voices and epistemologies, urging the inclusion of diverse perspectives in global climate conversations. Cherie Dimaline's *The Marrow Thieves* foregrounds the sacred interconnectedness of land and community, emphasizing the wisdom inherent in indigenous traditions while cautioning against techno-solutions that disregard ancestral knowledge and respect for nature.

By weaving together cultural histories and philosophies, world literature creates a pluralistic dialogue that decentralizes dominant narratives of climate engineering. It amplifies marginalized voices, encouraging alternative imaginaries rooted in coexistence and ecological harmony rather than control and dominance. Literature fosters a deeper understanding of how *climate engineering* impacts not only ecological systems but also societal structures, identities, and ethical responsibilities.

Through its exploration of speculative futures, diverse cultural narratives, and urgent ethical debates, world literature becomes an indispensable tool for imagining the complexities of an engineered Earth. It resists the reduction of environmental issues to purely scientific or technical terms, inviting readers to engage with *climate engineering* as a profoundly human and multifaceted phenomenon. By rethinking relationships between technology, justice, and nature, literature shapes not only public discourse but also the collective imagination needed to navigate the Anthropocene.

Many science fiction authors have explored *climate en-*

gineering in dystopian or speculative settings:

1. *The Ministry for the Future* by Kim Stanley Robinson

Kim Stanley Robinson's *The Ministry for the Future* stands as a towering achievement in climate fiction—a work that transcends the boundaries of traditional narrative to tackle one of the defining challenges of our time. At its core, the novel is a polyphonic narrative, meaning it interlaces multiple voices and viewpoints. From bureaucrats and scientists to activists and ordinary citizens, each character contributes a unique perspective on climate change. This narrative layering not only deepens the reader's engagement with the complex realities of the modern world but also mirrors the multifaceted nature of human experience in the era of global warming.

Although Robinson's work is steeped in the formal aesthetics of postmodernism, it is distinguished by a fervent political commitment. Drawing on the theoretical insights of Fredric Jameson, critics note that *The Ministry for the Future* is an exception to the cultural dominance of postmodernism. Rather than retreating into cynicism or endless irony—a common postmodern trope—Robinson projects an ambitious utopian vision. In his narrative, the catastrophic effects of climate change serve as a catalyst for reconfiguring societal and economic structures, suggesting that the seemingly insurmountable crises of late capitalism can indeed be met with collective innovation and shared responsibility.

At the same time, employing Mikhail Bakhtin's emphasis on polyphony, the novel celebrates the chorus of diverse voices. Each narrative strand, whether intimate or epic in scope, reminds us that solutions to global problems require a multiplicity of ideas and an open dialogue between communities. This approach aligns with Bakhtin's belief that a true conversation—one in which conflicting perspectives coexist—can foster a richer, more honest understanding of reality.

Robinson's text is more than a work of fiction—it is a pointed critique of the entanglements between climate change and the political and economic systems of our time. Through its depiction of *climate engineering* projects and innovative governance mechanisms, the novel dissects the power structures inherent in late capitalism. The narrative underscores how economic injustice, driven by the relentless pursuit of profit, not only accelerates environmental degradation but also deepens global inequality.

By foregrounding the role of ordinary citizens in mobilizing change, Robinson offers a vision of global cooperation where political will and grassroots activism converge. The narrative is replete with examples of collective action, from orchestrating radical economic reforms to engaging in policy debates and even employing ecological justice as a litmus test for new forms of corporate responsibility. In essence, the novel argues that when people work together on a global scale, they possess the power to transform not only their immediate surroundings but also the broader societal framework.

One of the most striking dimensions of *The Ministry for the Future* is its stark presentation of our current era as a crossroads—a choice between utopia and catastrophe. Drawing on utopian thinking, Robinson does not offer a simple escape into idealized futures; instead, he advocates for a "both/and" strategy. This perspective acknowledges the enormity of the crisis while simultaneously inspiring hope through community-based and democratic responses.

In this vision, green capitalism emerges as a viable alternative. Corporations are reconceived not as monolithic enemies of the environment but as potential stewards of a sustainable future. Yet, this reimagined capitalism is not a blanket endorsement of market forces. It is deeply imbued with ethical responsibility, demanding that economic activity should be subservient to the imperatives of ecological balance and social justice. In doing so, the novel challenges readers to rethink conventional economic models and embrace multifaceted strategies that combine civic engagement, government policy, and private sector adjustment.

The Ministry for the Future is a multifaceted narrative that unites speculative vision with rigorous political and ethical inquiry. It is a work that operates on several levels—offering both an immediate call to action and a long-term philosophical meditation on what it means to be human in a time of climatic upheaval. Through its polyphonic structure, its critical engagement with postmodern theories, and its resolute push for global cooperation and economic justice, the novel invites us to reimagine the future. It reminds us that in the face of overwhelming global challenges, the pathways to salvation lie in the diverse, often discordant voices of the people—a true testament to the power of literature to inspire, provoke, and ultimately transform society.

2. The Marrow Thieves by Cherie Dimaline

Cherie Dimaline's *The Marrow Thieves* (2017) is a powerful work of Indigenous speculative fiction that reclaims the future by rooting it in memory, land, and story. Set in a dystopian Canada ravaged by climate collapse, the novel imagines a future in which non-Indigenous people have lost the ability to dream. In response, state institutions begin harvesting the bone marrow of Indigenous peoples, who retain this essential human function—in a chilling revival of the logic of colonial extraction and assimilation.

Yet, rather than giving in to despair, Dimaline offers a narrative of resistance, resurgence, and relational ethics.

Through its structure, genre, and Indigenous worldviews, *The Marrow Thieves* transcends dystopia to assert an enduring future for Indigenous youth, culture, and land-based knowledge.

Positioned within the realm of Indigenous speculative fiction, *The Marrow Thieves* is also a prime example of what scholars term counter-genetic fiction—a genre that critiques the reduction of Indigenous identity to biological or genetic essentialism. In Dimaline's world, the commodification of Indigenous marrow reflects historical and ongoing attempts to exploit Indigenous bodies and cultures under settler colonialism.

However, the novel resists this through its embrace of story, memory, and intergenerational knowledge. These elements become the true carriers of identity, transcending any fixed genetic coding. In this way, the narrative insists that indigeneity is not something you inherit biologically but something you enact, remember, and live concerning others—including ancestors, land, and nonhuman beings.

One of the novel's most distinctive features is its nonlinear, spiraling narrative structure, which mirrors the Indigenous concept of time as non-hierarchical and cyclical. Through embedded oral stories, dreams, and personal recollections, Dimaline collapses the boundaries between past, present, and future.

These techniques reinforce the relational temporality that governs Indigenous storytelling, where the past informs the present, and the future is dreamed through collective memory. Dreams in the novel are not only metaphysical spaces of connection but also act as political acts of resistance, a reclaiming of cultural identity in the face of systemic erasure.

At the heart of *The Marrow Thieves* is the power of storytelling—both as a tool of survival and a weapon of resistance. The protagonist, Frenchie, and his companions navigate a fractured landscape not only by avoiding physical capture but by anchoring themselves in communal stories, oral traditions, and ceremonies.

This storytelling practice reinforces what Dimaline sees as essential to Indigenous resurgence: the ethics of reciprocal care. These ethics extend beyond the human, embracing nonhuman relations—animals, land, water—as kin. This more-than-human worldview challenges settler capitalist ideologies that view nature as property or resource, instead foregrounding relationships based on respect, balance, and responsibility.

Dimaline, herself Métis, brings to the novel a perspective shaped by the contemporary Métis condition, marked by displacement, dispossession, and cultural resilience. The narrative explicitly invokes the trauma of the Residential School System, where generations of Indigenous children were taken from their families, stripped of their languages and cultural practices, and often subjected to horrific abuse.

In The Marrow Thieves, the dystopian marrow-harvesting schools are a speculative reflection of real historical violence, yet they also serve to spotlight the resilience of Indigenous communities. By portraying dreams as sites of cultural memory and resistance, Dimaline transforms what was once targeted for erasure into a radical source of power and survival.

Though marketed as young adult (YA) fiction, *The Marrow Thieves* is deeply political. It centers Indigenous youth not as passive victims but as leaders of the future, inheritors of trauma who are also agents of healing and transformation. The YA format allows Dimaline to reach a broad and formative audience, offering them both an education in Indigenous history and a vision of solidarity, survival, and sovereignty.

By crafting characters who grow through communal care, mentorship, and shared memory, the novel becomes a manual for decolonial survival—a speculative but grounded vision of Indigenous resurgence amidst ecological and political collapse.

Cherie Dimaline's *The Marrow Thieves* is a searing and tender meditation on what it means to be Indigenous in the face of apocalypse—not just environmental but cultural and ontological. In reclaiming dreams as sacred, in privileging stories as repositories of survival, and in centering reciprocal relations with all beings, the novel challenges the extractive logics of settler colonialism and imagines alternative futures rooted in Indigenous ways of knowing.

In the growing canon of climate fiction and world literature, *The Marrow Thieves* stands out for its refusal to separate environmental justice from cultural survival. It reminds us that any conversation about climate change and sustainability must begin with a reckoning of colonial histories, and that the most radical acts of resistance may be found not in technology, but in remembering how to live in relation—to each other, to the land, and to the stories that sustain us.

3.Madd Addam Trilogy by Margaret Atwood.

Margaret Atwood's *Madd Addam Trilogy* is a visionary work of speculative fiction that interrogates the consequences of ecological collapse, corporate overreach, biotechnological obsession, and social disintegration. Composed of *Oryx and Crake* (2003), *The Year of the Flood* (2009), *and Madd Addam* (2013), the trilogy offers a layered and polyphonic narrative that moves between multiple characters, temporalities, and perspectives to provide a complex, intersectional critique of contemporary society.

Rather than being a cautionary tale steeped solely in

despair, Atwood's trilogy presents a nuanced exploration of dystopia and utopia, blending apocalyptic realism with mythic storytelling, ultimately pushing readers to contemplate ethical survival in the Anthropocene epoch.

The trilogy's world is shaped by late capitalist excess, ecological devastation, and biopolitical control. From corporate-controlled gated compounds to anarchic pleeblands, Atwood constructs a dystopia where social inequality, scientific hubris, and environmental degradation intertwine. The Waterless Flood—a man-made plague unleashed by the bioengineer Crake—is both a punishment and a reset, wiping out most of humanity.

Yet, amid the wreckage, Atwood plants the seeds of utopian renewal. The genetically modified Crakers, created by Crake to embody peace, simplicity, and ecological harmony, suggest a potential future not bound by traditional human flaws. Similarly, the God's Gardeners—eco-religious activists from *The Year of the Flood*—offer a spiritual and ecological counter-narrative rooted in care, interdependence, and nonviolence.

By refusing to fully commit to either dystopia or utopia, Atwood aligns with what Tom Moylan terms "critical dystopia": a genre that reveals the horrors of the present but gestures toward the possibility of resistance and renewal.

The trilogy's apocalyptic core, most evident in *Oryx and Crake,* is not a supernatural judgment but a product of human arrogance and techno-scientific overreach. The Waterless Flood, orchestrated under the guise of salvation, is both bioterrorism and misguided idealism. Through this narrative device, Atwood critiques not only the hubris of transhumanism but also the commodification of life itself.

However, the trilogy's polyphonic narrative structure—with voices like Snowman/Jimmy, Toby, Ren, Zeb, and the storytelling Crakers—undermines a single apocalyptic reading. These multiple perspectives foster a dialogic vision of survival, echoing Mikhail Bakhtin's concept of polyphony, where meaning emerges through dynamic interaction rather than singular authority.

Each character's account adds emotional and ideological texture to the post-apocalyptic world, offering both trauma and tenderness, memory and myth, despair and dreaming.

The trilogy is profoundly shaped by posthumanist concerns, questioning the boundaries between human and nonhuman, organic and synthetic, natural and artificial. The Crakers, pigoons (intelligent pigs bred for organ harvesting), and other bioforms populate a world where human exceptionalism has faltered.

Atwood does not romanticize these new beings but uses them to critique anthropocentrism and propose an alternative ethics of relationality, wherein survival depends on cooperation across species. The Crakers, in particular, represent a radical rethinking of human subjectivity—devoid of jealousy, greed, or hierarchy, their society reflects what a deconstructed, decentered humanity might look like.

This posthuman landscape also invites questions of ecological stewardship. Can beings not burdened by human history build a better world? Or are they still shaped by the legacy of their human creators?

A central feminist current runs through the trilogy, most vividly embodied by the character of Toby in *The Year of the Flood* and *Madd Addam*. In contrast to male figures like Jimmy or Crake, Toby exhibits a feminist new materialist ethos, grounded in care, attention to the more-than-human world, and non-dominating relationships.

Toby's bond with bees, her herbal knowledge, and her spiritual ecology position her as a figure of resistance against both patriarchal power and techno-capitalist instrumentalism. Atwood contrasts Toby's ethics of care with the masculinist, imperialist logic of domination embodied by Crake.

The trilogy also destabilizes binary gender norms, offering characters like Amanda and Ren, who refuse to be categorized within traditional gender expectations. Moreover, Atwood integrates insights from anthropocene feminism, challenging the universality of "man" and foregrounding interconnected vulnerabilities—of species, ecosystems, and marginalized bodies.

The *Madd Addam* world is shaped by neoliberal biopolitics—a system that commodifies bodies, genes, and even the apocalypse itself. The trilogy draws connections between contemporary corporate capitalism and colonialism, portraying how bodies (especially women's and animal bodies) are mined, mutated, and marketed.

Atwood also stages the "end of history" in multiple dimensions: the exhaustion of Enlightenment narratives of progress, the flattening of historical consciousness by corporate media, and the literal end of the human species as we know it. Yet by crafting an alternative future grounded in memory, oral storytelling, and ecological wisdom, Atwood reopens history as a space of possibility.

Margaret Atwood's *Madd Addam Trilogy* is a landmark in speculative fiction, offering an incisive and layered critique of ecological collapse, capitalist overreach, and gendered domination. Yet it also gestures toward the potential for new forms of life, subjectivity, and community beyond the limits of the human.

Through its polyphonic voices, posthuman visions, and feminist ethics, the trilogy affirms that even in the ashes of the Anthropocene, there can be stories of survival, memory, care, and resistance. Not a utopia, not quite a dystopia, Madd Addam envisions a world where the end is only a beginning—if we are willing to listen, adapt, and live differently.

Narrating Engineered Futures: Literature as Witness and Ethical Compass

In the face of escalating climate crises and the controversial rise of *climate engineering* as a proposed solution, literature in the Anthropocene assumes a critical, imaginative, and ethical role. It interrogates the technocratic logic that underpins geoengineering strategies, revealing the deep entanglements between environmental degradation, power structures, and human hubris. Literary narratives do not merely reflect ecological anxieties; they reframe them, offering speculative visions, moral counterpoints, and voices for the silenced, both human and nonhuman.

By exploring the implications of manipulating Earth's systems, literature resists reductive narratives of control and mastery, instead fostering a planetary consciousness rooted in humility, interdependence, and care. In doing so, it contributes to a more nuanced discourse that challenges the dominance of scientific determinism and opens ethical deliberations about justice, responsibility, and survival. As the Anthropocene unfolds, literature stands as both witness and participant, bearing stories of caution, resistance, and hope in the face of engineered futures.

CHAPTER 5

Grief in the *Anthropocene*: Environmental Loss and Emotional Landscapes in Climate Novels.

Ecological grief refers to the emotional distress and mourning individuals or communities undergo due to environmental loss. This phenomenon has gained increased scholarly and popular attention in recent years, especially concerning climate change and environmental disasters. It encompasses a spectrum of psychological responses—from sadness and helplessness to existential anxiety—arising from the degradation or anticipated destruction of ecosystems, biodiversity, and landscapes that hold cultural, spiritual, or personal significance (Cunsolo & Ellis, 2018).

The concept finds early articulation in Aldo Leopold's *A Sand County Almanac* (1949), where he expressed the pain of witnessing ecological degradation and emphasized the ethical responsibility of humans toward the natural world. Although Leopold did not formally coin the term "*ecological grief*," his reflections laid the groundwork for later theorists. The term gained academic legitimacy in the 21st century, particularly through the work of Ashlee Cunsolo and Neville Ellis. Their research documented how communities, particularly Indigenous populations and rural societies, experience grief due to climate-related disruptions, biodiversity loss, and the erosion of traditional ecological knowledge (Cunsolo & Ellis, 2018).

Recognizing *ecological grief* is crucial for understanding the full impact of environmental crises—not merely in physical or economic terms but also in psychological and emotional dimensions. Scholars have identified three primary manifestations of this grief. First, the loss of place and landscape, where communities are physically displaced or environments are radically altered. Second, the loss of cultural identity, where traditional practices rooted in specific ecosystems are eroded. Third, anticipatory grief, characterized by emotional turmoil over impending or inevitable ecological collapse (Ojala, 2017).

As environmental degradation intensifies, literature has emerged as a vital medium for articulating and processing *ecological grief.* Through speculative, dystopian, and realistic narratives, literature provides a space to explore the affective dimensions of environmental change and imagine alternative futures. This essay critically examines how four notable novels—Richard Powers' The Overstory (2018), Paolo Bacigalupi's *The Water Knife* (2015) and *The Wind-up Girl* (2009), and Ursula K. Le Guin's *The Dispossessed* (1974)—address *ecological grief,* offer commentary on environmental ethics, and promote ecological awareness.

1. *The Overstory* by Richard Powers: Grieving the Silencing of Trees.

Richard Powers' The Overstory is a polyphonic and sprawling narrative that interweaves the lives of nine characters, each connected through their profound experiences with trees. The novel operates as a literary homage to arboreal life, detailing the intelligence, interconnectedness, and resilience of trees in the face of human destruction. Powers invokes *ecological grief* most explicitly in the character arcs of Patricia Westerford, an ecologist whose work on tree communication is dismissed by the scientific community, and Olivia Vandergriff, whose spiritual transformation after a near-death experience leads her to eco-activism.

"The Overstory" is a sweeping novel that explores the lives of the characters connected by their relationships with trees. The characters in "The Overstory" are complex and multifaceted, with each one bringing their own unique perspective to the novel.

The novel explores the theme of *ecological grief* through the characters' experiences with deforestation, urbanization, and the destruction of natural habitats. The author masterfully conveys the emotional toll of environmental degradation, encouraging readers to reflect on their connection with nature. The novel highlights the importance of preserving the natural world and the need for sustainable practices.

The *ecological grief* in The Overstory emerges from both personal and collective mourning. Deforestation, habitat fragmentation, and biodiversity loss are portrayed not merely as environmental issues but as emotional catastrophes that destabilize identity and meaning. Patricia's observation that "trees are social creatures" (Powers, 2018, p. 203) challenges anthropocentric paradigms and urges readers to consider the silent suffering of the natural world. By foregrounding the emotional bond between humans and non-human life, the novel compels readers to grieve not just environmental loss but also humanity's failure to recognize ecological interconnectedness.

2. *The Water Knife by Paolo* Bacigalupi: Dystopia as Environmental Mourning

In *The Water Knife*, Paolo Bacigalupi envisions a near-future American Southwest devastated by water scarcity and climate-induced migration. The novel centers on Angel

Velasquez, a corporate enforcer protecting water rights for wealthy enclaves, and Lucy Monroe, a journalist documenting the human cost of ecological collapse. Bacigalupi's vision of a fragmented, violent, and resource-stricken society serves as a bleak warning about the future of environmental mismanagement.

The novel is set in a dystopian future world where climate change has ravaged the environment, leading to widespread water scarcity. The setting is vividly described, and the author uses the natural world to explore themes of ecology, conservation, and the human relationship with nature.

Ecological grief in *The Water Knife* is palpable in the desperation of displaced populations and the commodification of natural resources. Unlike Powers' elegiac tone, Bacigalupi's narrative is gritty and instinctive, exposing the systemic inequalities exacerbated by environmental collapse. The grief here is twofold: a lament for lost landscapes and a cry against the moral bankruptcy of institutions that prioritize profit over people. Bacigalupi's dystopia is not only a cautionary tale but a meditation on how climate change dehumanizes and desensitizes societies, leading to what Cunsolo and Landman (2017) term "disenfranchised grief"—a sorrow that is ignored or invalidated by dominant systems of power.

3. *The Windup Girl* by Paolo Bacigalupi: Technological Hubris and Ecological Sorrow.

Paolo Bacigalupi's *The Windup Girl* (2009) is a landmark work in climate fiction (cli-fi) that deeply engages with the theme of *ecological grief*—a profound mourning for environmental degradation and irreversible ecological loss. Set in a future Thailand devastated by biotechnological warfare, rising sea levels, and genetic collapse, the novel portrays a society teetering on the edge of ecological and cultural extinction. Bacigalupi presents a world where grief is not just personal but collective, embedded in the very structure of daily life and political power.

Ecological grief manifests in the fear and resentment toward genetically modified organisms and the corporatization of seeds and food. Characters like Anderson Lake, an agent of an agro-corporation, exploit this grief, seeking to control Thailand's seedbank—one of the few remnants of unmodified biodiversity (Bacigalupi, 2009). Meanwhile, the environmental ministry's strict preservationist policies reflect a desperate attempt to mourn and protect what little is left of a once-abundant natural world. This mourning is not passive; it fuels political resistance and violence, underscoring how grief can become a force of ecological and social agency.

Emiko, the titular "windup girl," a genetically engineered being, embodies a more intimate form of *ecological grief*. Her very existence reflects a world that has commodified life, severed from natural reproduction and connection. Her alienation is not just social but existential—she is a symbol of ecological displacement and the grief of being out of place in a mechanized, ruined biosphere.

Bacigalupi's narrative does not offer hope in the traditional sense but instead forces readers to confront the consequences of unchecked ecological abuse and corporate greed. The city of Bangkok, surrounded by walls against the rising sea, becomes a metaphor for humanity's futile attempts to stave off the inevitable consequences of its own actions (Bacigalupi, 2009). In this way, *The Windup Girl* does not merely illustrate environmental decay but mourns it, transforming the novel into a eulogy for lost ecologies and a meditation on the emotional toll of planetary collapse.

4. *The Dispossessed* by Ursula K. Le Guin: Utopian Resistance to *Ecological Grief*.

Contrasting with the dystopian despair of Bacigalupi

and the lamentation in Powers, Ursula K. Le Guin's *The Dispossessed* offers a vision of ecological harmony and ethical coexistence. Set between the anarchist planet Anarres and the capitalist Urras, the novel follows Shevek, a physicist navigating the moral and philosophical contradictions of both worlds. Anarres, though barren and resource-poor, sustains a society rooted in mutual aid, environmental responsibility, and egalitarianism.

The novel portrays the consequences of human actions on the environment, particularly on the planet Urras. The degradation of the environment is depicted because of the capitalist society's focus on growth and profit. This degradation has a profound impact on the characters and their relationships with the natural world.

The novel's portrayal of environmental degradation and its impact on human relationships can be seen as a form of *ecological grief*. The characters' experiences and emotions are deeply connected to the natural world, and the destruction of the environment has a profound impact on their well-being.

Ecological grief in The Dispossessed is less overt and more anticipatory. It exists as a cautionary undercurrent—a sense that the sustainable values of Anarres could be lost if the society succumbs to hierarchy and competition. Le Guin's novel envisions a path not of mourning but of prevention, asserting that alternative political and ecological structures can mitigate the grief caused by environmental decline. In doing so, Le Guin aligns with eco-utopian traditions, where literature becomes a blueprint for resilience and hope rather than despair (Buell, 2005).

Literature as a Catalyst for Climate Consciousness

Each of these novels employs *ecological grief* as a narrative device to foster environmental awareness. Through their characters and worlds, they illuminate the psychological impacts of ecological devastation, critique unsustainable systems, and offer models—whether through protest, reflection, or utopian vision—for coping with loss.

They depict characters in mourning over environmental changes, these novels validate *ecological grief* as a legitimate and necessary response to climate change. Bacigalupi's works expose the structural causes of environmental collapse—capitalism, privatization, and technological hubris—while Powers and Le Guin emphasize relational ethics and ecological literacy. While some novels dwell in dystopian realities, others, particularly The Dispossessed, offer glimpses of alternative societies built on ecological stewardship.

By engaging readers in emotional, moral, and philosophical dialogues, these literary works function as more than cautionary tales—they become tools for consciousness-raising and collective healing in the *Anthropocene.*

Ecological grief is a multifaceted emotional condition that reflects the profound human connection to the environment. As climate change accelerates, literature becomes an essential medium for exploring this grief, offering narratives that validate sorrow, critique destructive paradigms, and imagine hopeful alternatives. The four novels examined—The Overstory, *The Water Knife*, *The Windup Girl*, and The Dispossessed—demonstrate how fiction can illuminate the psychological toll of environmental change and inspire renewed commitments to ecological care. Through the lens of *ecological grief*, literature invites us to feel, reflect, and act.

CHAPTER 6

Eco-Anxiety: The Psychological Toll of the Climate Crisis

As climate change accelerates, its consequences extend beyond physical environmental damage into the psychological and emotional realms of human life. One of the most pervasive emotional responses to this global crisis is eco-anxiety—a chronic, often debilitating fear of ecological collapse. The American Psychological Association defines eco-anxiety as persistent worry, grief, and guilt related to environmental destruction, particularly climate change. While not classified as a clinical disorder, eco-anxiety has emerged as a significant mental health concern, especially among youth.

In this context, world literature serves as both a mirror and a mediator of collective ecological consciousness. Literature across cultures increasingly engages with themes of environmental degradation, displacement, and psychological distress, providing a space for emotional processing, ethical reflection, and sociopolitical critique. This paper explores how world literature represents and responds to eco-anxiety and how literary narratives help individuals and societies cope with the realities of environmental change.

Eco-anxiety is characterized by persistent feelings of fear, guilt, grief, and helplessness stemming from real or perceived threats to the environment. It arises from increased awareness of global ecological crises such as rising sea levels, wildfires, biodiversity loss, and extreme weather events. According to recent psychological studies, eco-anxiety affects both individual mental health and collective emotional states, manifesting in stress-related symptoms, existential dread, and behavioral changes.

The American Psychological Association underscores the growing prevalence of eco-anxiety, particularly among young people and women. Cultural and contextual factors shape how eco-anxiety is experienced, with those in vulnerable ecological zones or socioeconomically disadvantaged communities reporting heightened emotional responses.

While the term "eco-anxiety" is contemporary, literature has long grappled with human responses to nature's power and fragility. From Romantic poets lamenting industrialization to postcolonial ecocriticism highlighting environmental exploitation, writers have historically responded to ecological change with dread and reverence.

Cultural narratives influence how communities interpret and express ecological fear. In Indigenous literature, for instance, eco-anxiety often intersects with themes of dispossession and spiritual loss. In contrast, Western texts may foreground scientific rationalism and individual ethical responsibility. These diverse literary approaches reveal that eco-anxiety is not a monolithic experience but is shaped by geography, culture, and historical memory.

One of the primary causes of eco-anxiety is exposure to climate disasters. Individuals who have lived through devastating events like hurricanes, wildfires, floods, or droughts often face long-term emotional distress. For example, wildfire survivors in California and Australia have reported ongoing trauma linked to the loss of their homes, communities, and sense of safety. Another significant contributor is constant media exposure and climate warnings. The relentless stream of alarming news, such as reports about the Amazon rainforest burning or the rapid death of coral reefs, can foster a pervasive sense of doom and helplessness. Adding to this psychological burden is political inaction and climate denial. The apparent unwillingness of governments and corporations to implement meaningful climate policies often leaves citizens, especially young activists like Greta Thunberg, feeling frustrated and powerless. Eco-anxiety is also intensified by guilt over personal carbon footprints as individuals struggle with their roles in perpetuating environmentally harmful systems. The challenge of living a truly sustainable lifestyle can be overwhelming and emotionally draining. Lastly, the loss of natural spaces and wildlife can lead to a form of grief known as *ecological grief*. Scientists who witness the bleaching of coral reefs, for instance, frequently report profound sadness and anxiety over the irreversible damage to Earth's ecosystems.

World literature provides a dynamic platform for expressing eco-anxiety. Novels, poems, and speculative fiction reflect and shape collective ecological consciousness:

1. *Generation Dread* by Britt Wray

Britt Wray's *Generation Dread: Finding Purpose in an Age of Climate Crisis* is a landmark contribution to the emerging field of climate psychology. It offers an emotionally resonant and intellectually rigorous account of eco-anxiety—a phenomenon now widely acknowledged as an urgent mental health issue in the context of escalating climate disruption. Through a blend of memoir, scientific research, and cultural critique, Wray delivers a compassionate and pragmatic exploration of how environmental degradation impacts not only the physical world but also the inner landscapes of human beings.

A central thesis of *Generation Dread* is that eco-anxiety should not be pathologized but rather recognized as a sign of emotional and ecological intelligence. Wray asserts that experiencing anxiety, grief, and dread in the face of environmental collapse is a rational and empathetic response. She reframes these emotions as necessary catalysts for climate engagement, arguing that suppressing or dismissing them only deepens feelings of alienation. This perspective shifts the focus from emotional paralysis to transformative action, encouraging individuals to embrace their vulnerability as a source of strength.

One of the book's most commendable contributions lies in its intersectional approach. Wray meticulously unpacks how climate anxiety is unevenly experienced across lines of race, class, gender, geography, and generational status. By spotlighting the voices of Indigenous thinkers, activists from the Global South, and people of color, she broadens the discussion of eco-anxiety beyond the white, Western, middle-class demographic that often dominates environmental discourse. In doing so, *Generation Dread* not only deconstructs privilege within climate narratives but also calls for inclusive, equitable solutions that foreground environmental justice.

Wray ventures into largely uncharted territory by advocating for climate-informed mental health care. She interviews psychologists and therapists who are pioneering new frameworks to support those grappling with climate grief, ecological despair, and anticipatory trauma. The book emphasizes the need for therapeutic spaces where individuals can safely explore their environmental fears without being labeled alarmist or irrational. Wray positions therapy not as a means to silence anxiety but as a tool to metabolize it, making room for creativity, resilience, and community action.

Compassion emerges as both a theme and a method in Generation Dread. Wray encourages readers to channel their emotional responses into acts of care toward the planet, one another, and themselves. She discusses how cultivating compassion can bridge generational divides, deepen political solidarity, and inspire collective resilience. In a time when the climate crisis can provoke nihilism or disengagement, Wray's vision of care-based activism is both radical and necessary.

Stylistically, Wray strikes a rare balance between accessibility and academic integrity. Her writing is deeply personal yet rooted in scholarly research, making *Generation Dread* a valuable resource for mental health professionals, environmental educators, students, and general readers alike. Her voice is at once scientific and soulful, capable of speaking to the intellectual, emotional, and ethical dimensions of the climate crisis.

Critics and readers alike have lauded *Generation Dread* for its courage, clarity, and compassion. It has been praised by public intellectuals such as Naomi Klein for its bold confrontation of the psychological costs of climate change, while Dr. Gabor Maté commends its insights into trauma and healing. The book's nomination as a finalist for the Governor General's Literary Award and its inclusion among the best Canadian nonfiction books of 2022 further affirm its significance.

Generation Dread stands as a seminal work in contemporary climate literature. Britt Wray does not offer false reassurance or quick solutions; instead, she provides the tools to live meaningfully and ethically within an age of uncertainty. Her message is clear: in the face of climate dread, our best hope lies not in denial but in connection—connection to our feelings, to each other, and to a planet in peril. Through this connection, we can begin to imagine—and build—a more compassionate and sustainable future.

2. ***Eco-Anxiety and Environmental Education*** **by Panu Pihkala.**

In Eco-Anxiety and Environmental Education (2020), Finnish scholar and theologian Panu Pihkala delivers a pioneering exploration of the intersection between environmental education and the growing psychological phenomenon of eco-anxiety. This work stands as one of the earliest comprehensive academic efforts to address eco-anxiety not as a fringe concern but as a central emotional and educational challenge in the Anthropocene era. Pihkala's unique interdisciplinary lens—drawing on theology, psychology, education, and environmental studies—allows him to present eco-anxiety as a complex, culturally situated, and deeply human response to the climate crisis.

Pihkala defines eco-anxiety as a broad spectrum of emotional responses to ecological crises, encompassing grief, helplessness, fear, guilt, and existential dread. He emphasizes that these reactions are not pathological but rather deeply reasonable emotional responses to genuine global threats. By situating eco-anxiety in historical and cultural contexts, he pushes readers to consider how such emotions have always accompanied shifts in human-environment relationships, though now intensified by the rapid acceleration of climate change and biodiversity loss.

Importantly, the book expands the concept of eco-anxiety beyond clinical psychology, arguing that it must be addressed in educational frameworks, particularly those that shape the worldviews of younger generations. Pihkala advocates for a new kind of environmental education—one that not only informs but also emotionally equips learners to process environmental loss and uncertainty.

One of Pihkala's central arguments is that environmental education must acknowledge eco-anxiety if it is to remain relevant and humane. He critiques traditional models that focus solely on disseminating ecological facts, often leading to emotional overwhelm or paralysis among students. Instead, he calls for an emotionally literate pedagogy that integrates psychological support, ethical reflection, and experiential connection with nature.

He presents nature connection, narrative therapy, mindfulness, and ecotherapy as essential components of an education that fosters not only ecological literacy but also emotional resilience. Through this approach, students can navigate the emotional burdens of environmental knowledge without succumbing to despair.

Pihkala devotes significant attention to the psychological vulnerability of children and adolescents, for whom eco-anxiety can manifest more intensely due to limited coping mechanisms and the burden of inheriting an uncertain future. He explores how environmental trauma, such as witnessing ecosystem loss or experiencing climate-related disasters, can have long-term developmental consequences.

He also addresses the ethical dilemma faced by educators: how to inform without traumatizing. Pihkala recommends age-appropriate engagement, emphasizing hope, empowerment, and collective action rather than fear-based messaging.

Pihkala is sensitive to socioeconomic and cultural differences in the experience of eco-anxiety. He notes that while the phenomenon has been widely discussed in the Global North, communities in the Global South often experience direct environmental trauma without the luxury of conceptualizing it through psychological terms. His analysis of intersectionality reveals how climate-induced emotional distress intersects with issues of race, class, and historical environmental injustice.

Another critical insight in the book is the role of media in amplifying eco-anxiety. While increased visibility of environmental issues can promote awareness, it can also fuel despair, particularly in younger generations who are constantly exposed to ecological devastation via digital platforms. Pihkala urges for a more ethical media discourse—one that balances urgency with constructive narratives and emotional validation.

Pihkala reviews a range of therapeutic approaches—including psychoanalytic, grief-focused, cognitive behavioral, acceptance and commitment, and narrative therapies—to address eco-anxiety. He also emphasizes the potential of art-based interventions, group support, and storytelling in processing ecological emotions.

Moreover, he argues for systemic change in education, proposing that curricula at all levels incorporate nature-related practices, emotional expression, and reflective dialogue. Only by attending to the emotional dimensions of environmental crises can education prepare individuals to face the Anthropocene with courage and care.

Eco-Anxiety and Environmental Education is a seminal text that elevates eco-anxiety from a personal affliction to a collective educational and ethical concern. Pihkala's work is essential reading for educators, psychologists, policymakers, and environmental advocates. His compassionate, nuanced, and interdisciplinary approach offers a framework not only for understanding the psychological dimensions of the climate crisis but also for responding to them with empathy and insight.

At a time when many educational systems are struggling to adapt to the emotional realities of environmental collapse, Pihkala's vision of hopeful, holistic, and emotionally engaged education is both timely and transformative.

3. *Active Hope: How to Face the Mess We're in with Resilience & Creative Power* by Joanna Macy and Chris Johnstone.

In *Active Hope,* Joanna Macy and Chris Johnstone offer a transformative approach to confronting the daunting ecological and social crises of our time. Rather than framing hope as a passive emotional state, the authors recast it as an active practice, rooted in intention and empowered by community, resilience, and a deep sense of interconnectedness. This book emerges not merely as a response to the climate emergency and ecological degradation but as a guide for navigating the emotional, psychological, and spiritual terrain of life in the Anthropocene.

Joanna Macy is a leading voice in eco-philosophy, known for her pioneering contributions to deep ecology, Buddhism, and systems theory. Her decades-long engagement with social and ecological movements inform her interdisciplinary approach to activism and healing. Macy has developed a unique methodology known as the Work That Reconnects, a workshop-based framework for emotional processing and collective empowerment in the face of global crises. Her co-authorship of *Active Hope* with medical doctor and resilience specialist Chris Johnstone offers a synthesis of psychological insight and spiritual depth, making the book both accessible and profound.

This spiral is not just metaphorical but offers a practical, process-oriented structure that allows individuals and communities to move through despair, grief, and overwhelm toward renewed agency and commitment. Macy and Johnstone argue that despair is not a weakness but a gateway to transformation, a sign that we are still emotionally invested in the world.

The authors introduce the concept of "*active hope*" as a practice of consciously choosing the future we want to work toward, even when outcomes remain uncertain. This kind of hope is not based on wishful thinking but on commitment and participation. Drawing from psychological research and spiritual traditions, they demonstrate how hope functions as a survival strategy, deeply tied to our capacity for resilience, creativity, and community connection.

Hope, they argue, enhances resilience by helping people interpret adversity in ways that are empowering rather than debilitating. It allows individuals to remain motivated, find meaning, and persist through challenging times. Numerous psychological studies support this view, showing that hope accounts for a significant percentage of the variance in resilience. In this light, hope is not an escape from reality but an anchor that holds us steady amid uncertainty.

The book is especially relevant in the context of the Anthropocene, an epoch marked by human-driven environmental change. Macy and Johnstone address the psychological toll of this era—feelings of being overwhelmed, grief, guilt, and helplessness—through tools such as group work, mindfulness, storytelling, and systems thinking. Their methodology encourages individuals to transform personal anguish into collective strength, emphasizing the interconnectedness of all life and the power of community in sustaining meaningful action.

Active Hope has found resonance across a wide range of disciplines—from environmental education and psychology to spiritual care and grassroots activism. Its influence has grown in the wake of crises such as COVID-19, where individuals have been forced to confront the fragility of life systems and the importance of inner resilience.

The book's teachings have been incorporated into educational programs, therapy models, and activist circles, making it more than a theoretical work—it is a lived philosophy. The emphasis on participatory hope and ecological consciousness offers a blueprint for individuals seeking to make sense of, and act within, the world's complex ecological challenges.

In *Active Hope*, Macy and Johnstone deliver a work that is as emotionally restorative as it is strategically empowering. It does not shy away from the harsh truths of our environmental reality but teaches us to face them with open hearts and active hands. As despair, grief, and eco-anxiety become increasingly common emotional states in a rapidly destabilizing world, this book offers both a mirror and a lantern—reflecting our pain while illuminating the path forward.

For anyone seeking guidance on how to remain whole, grounded, and effective in the face of planetary crises, Active Hope is not merely recommended reading—it is essential.

From Dread to Dialogue: Literature's Role in Navigating Eco-Anxiety

Eco-anxiety also drives behavioral responses, including environmental activism, changes in consumption, and shifts in lifestyle. However, it may also result in emotional paralysis, depression, or withdrawal. University students, for example, often report high levels of eco-anxiety, which correlate with impaired academic and social functioning.

In literature, this dual nature of eco-anxiety is explored through characters who oscillate between hope and hopelessness, action and apathy. Literature thus mirrors real-life coping strategies while offering new models of ecological consciousness and community.

Eco-anxiety is a profound psychological response to ecological crises that threaten not only individual well-being but also collective futures. As a reflection of human emotion and cultural consciousness, world literature plays a crucial role in exploring, validating, and transforming eco-anxiety. By engaging with diverse literary narratives, readers can gain insight into the emotional dimensions of climate change and discover pathways toward resilience, solidarity, and ecological hope.

CHAPTER 7

Writing Through the Wreckage: *Deep Adaptation* Narratives in Cli-Fi.

In the realm of climate fiction—often referred to as "cli-fi"—*Deep Adaptation* has emerged as a powerful literary lens for exploring how humans might respond to the irreversible consequences of climate change. This concept shifts the focus from prevention and mitigation to confronting the reality that some environmental changes, such as rising sea levels, persistent droughts, and intensified natural disasters, can no longer be avoided. *Deep Adaptation* encourages radical rethinking of how societies live, relate, and endure in the face of ecological breakdown (Bendell).

The concept of *Deep Adaptation* was formally introduced by British scholar Jem Bendell in his 2018 paper *Deep Adaptation: A Map for Navigating Climate Tragedy*. Bendell argued that humanity must now prepare not only for physical changes but also for profound psychological and cultural shifts brought on by climate collapse. Though American author Roy Scranton had earlier touched on similar themes in his work Learning to Die in the Anthropocene, it was Bendell who coined the term "*Deep Adaptation*" and framed it as a call for systemic transformation (Bendell; Scranton).

In literature, this concept has found a powerful outlet in climate fiction. Cli-fi not only imagines the future consequences of global warming but also dramatizes the emotional, societal, and ethical challenges of surviving in a transformed world. These narratives invite readers to reflect on what it truly means to adapt, not just technologically, but philosophically and communally.

Deep Adaptation literature is characterized by key themes that reflect humanity's evolving relationship with a changing planet. Central to these narratives is the idea of transformation and resilience, where characters must relinquish familiar comforts and adapt emotionally, morally, and physically to new ways of living. These stories often depict communities facing climate crises, highlighting how such events can either fracture societies or inspire grassroots cooperation and social change. Environmental justice emerges as a prominent concern, with many works exposing how climate change disproportionately affects marginalized populations, emphasizing the interconnectedness of inequality, race, and ecology, as noted by Nixon. Importantly, *Deep Adaptation* literature is not limited to dystopian visions; it also imagines alternative futures—speculative utopias or hybrid societies—where humans have found ways to live in harmony with nature, offering hope amid ecological despair.

The literary engagement with *Deep Adaptation* has sparked global awareness by making the climate crisis emotionally resonant and ethically urgent. Literature reaches people beyond academic or policy spheres. It provokes empathy, raises moral questions, and allows readers to rehearse responses to real-world climate scenarios. In this way, climate fiction contributes to cultural readiness, fostering the emotional intelligence and social imagination necessary for enduring environmental upheaval (Trexler).

This essay explores how some major works of climate fiction- *Oryx and Crake (2003), Parable of the Sower (1993), Suth Pole Station (2017), and Flight Behaviour (2012)*- engage with climate issues through the lens of *deep adaptation.* These texts not only illustrate the consequences of ecological neglect but also function as imaginative blueprints for survival, resistance, and hope in the face of planetary transformation.

1.Margaret Atwood's *Oryx and Crake:* A Vision of *Deep Adaptation* and Environmental Reckoning.

Margaret Atwood's *Oryx and Crake* (2003) is a striking example of climate fiction that powerfully imagines a future shaped by ecological collapse, biotechnological hubris, and moral ambiguity. Through its dystopian narrative, the novel aligns with the principles of *deep adaptation*, a framework that encourages humanity to acknowledge the possibility of inevitable environmental breakdown and to prepare for transformative change in societal values, structures, and relationships with nature.

The story unfolds in the aftermath of a global catastrophe, focusing on the character of Snowman, formerly Jimmy, who reflects on the downfall of human civilization. His memories of Crake, a brilliant yet dangerous bioengineer, and Oryx, a mysterious figure from his past, offer insight into a world where scientific experimentation and consumer culture pushed the Earth beyond repair. Crake's creation of the Crakers—a genetically engineered, docile humanoid species designed to replace humans—illustrates a radical, though ethically troubling, attempt to "solve" humanity's destructive tendencies by erasing them entirely.

Rather than offering a simplistic warning about climate change, Atwood's narrative pushes readers to consider the deep philosophical implications of environmental degradation. The novel poses difficult questions about what must be sacrificed, preserved, or reimagined in the face of ecological collapse. These themes echo Jem Bendell's "Four Rs" of *deep adaptation*: resilience, relinquishment, restoration, and reconciliation (Bendell, 2018). Snowman's struggle to survive in the ruins of a once-thriving civilization reflects a painful process of adaptation, not just to physical realities but to emotional and ethical disorientation.

The novel resonates globally by provoking critical reflection on human behaviour and its long-term consequences. Atwood's speculative world is not far removed from current ecological anxieties. Her writing, both vivid and unsettling, forces readers to confront the limitations of technological "solutions" and the urgency of shifting toward a more ecologically mindful and ethically grounded worldview. As environmental scholar Ursula Heise notes, Atwood "makes us rethink not only our past mistakes but also the kinds of futures we consider possible" (Heise, 2008, p. 125).

In essence, *Oryx and Crake* challenges audiences worldwide to engage with climate change not only through policy or innovation but through a deep, transformative reckoning with the systems and values that have brought humanity to this precarious moment. It is a novel that encourages resilience not just in surviving disaster—but in reimagining what it means to live responsibly on a shared planet.

2. Octavia E. Butler's *Parable of the Sower: Deep Adaptation* and Environmental Consciousness.

Octavia E. Butler's *Parable of the Sower* (1993) offers a prophetic and unsettling vision of a near-future America unravelling under the pressures of climate change, social collapse, and moral disintegration. Through the journey of its protagonist, Lauren Olamina, the novel articulates themes closely aligned with the principles of *deep adaptation*—a framework that urges humanity to accept and respond to the likely disruptions of the climate crisis not just with technological fixes, but through psychological, cultural, and ethical transformation.

Set in the 2020s, *Parable of the Sower* imagines a society where water is scarce, ecosystems have collapsed, and government institutions are failing. Rather than focusing on technological salvation, Butler presents a grounded, human-centered response to crisis: Lauren's creation of Earthseed, a belief system that emphasizes change, resilience, and adaptation, speaks to the core of what *deep adaptation* demands. Her mantra— "God is Change"—invites readers to view adaptation not as a temporary solution but as an enduring way of life. This mirrors Jem Bendell's *deep adaptation* agenda, which includes resilience, relinquishment, restoration, and reconciliation (Bendell, 2018), all of which are woven into Lauren's worldview and survival strategies.

Butler's novel does not romanticize collapse. Instead, it confronts its readers with a harsh yet honest depiction of what climate-induced breakdown could entail—violence, migration, inequality, and ecological loss. What makes the book resonate globally is its emphasis on individual agency, community-building, and ethical clarity in the face of systemic failure. Readers are compelled to reflect not only on the urgency of environmental action but also on the values and relationships we must cultivate in the process of adaptation.

In this way, *Parable of the Sower* contributes significantly to environmental discourse by illustrating that preserving nature and combating climate change is not only a matter of science or policy, but also of belief, resilience, and collective imagination. As scholar Amanda B. Baugh notes, Butler "uses speculative fiction to highlight environmental justice issues and to propose alternative ways of knowing and living with the Earth" (Baugh, 2017). Her work serves as a profound reminder that meaningful environmental change begins with a willingness to reimagine human priorities.

Ultimately, *Parable of the Sower* offers a powerful narrative of ecological and social transformation that resonates with global readers. It reinforces the importance of not only preventing climate catastrophe but also learning how to live meaningfully and ethically in a world shaped by it.

3. Ashley Shelby's *South Pole Station*: *Deep Adaptation* and the Art of Climate Awareness.

Ashley Shelby's *South Pole Station* (2017) is a satirical yet deeply insightful novel that places climate change at the heart of its narrative, not through apocalyptic spectacle but via human behaviour, institutional contradictions, and the fragile ecosystems of the Antarctic. With both humour and poignancy, Shelby explores how individuals and scientific communities confront (or deny) ecological realities, framing a subtle but profound commentary on the necessity of *deep adaptation* in the face of climate instability.

The novel follows Cooper Gosling, an artist who wins a fellowship to live and work at the Amundsen-Scott *South Pole Station.* Set in one of the most extreme and climate-sensitive regions on Earth, the story uses its isolated Antarctic setting to reflect on global environmental anxiety. The station, filled with scientists, climate deniers, bureaucrats, and eccentrics, becomes a microcosm of larger societal debates about science, truth, and survival. Within this melting pot of intellect and absurdity, Shelby illuminates how climate change is not only a scientific issue but also a deeply cultural and existential one.

South Pole Station resonates with the principles of *deep adaptation*, particularly in its treatment of emotional and institutional responses to environmental collapse. As Jem Bendell (2018) outlines, *deep adaptation* involves preparing for the social and psychological effects of climate breakdown, including questioning core values, adjusting expectations, and finding new ways to live meaningfully in a disrupted world. Shelby captures this through Cooper's personal transformation—her gradual confrontation with grief, confusion, and the need for purpose in a world on the brink.

Rather than offering easy solutions, Shelby's novel encourages readers to reflect on the ways institutions—scientific, governmental, and personal—respond to crisis, often with denial, inertia, or satire. Yet, through Cooper's evolving relationships and artistic expression, *South Pole Station* also presents a path toward reconciliation, curiosity, and courage in the face of uncertainty. The novel ultimately asks: how do we live with knowledge of the Earth's decline? What can art, science, and human connection offer us in a time of planetary distress?

The story's humour does not dilute its message; rather, it makes the global urgency of climate change more accessible and relatable. Readers around the world can identify with the characters' confusion, fear, and resilience, making *South Pole Station* a quietly powerful appeal for climate responsibility and emotional preparedness. As eco-critic Adam Trexler notes, "climate fiction often explores not just environmental change, but the psychological reckoning that accompanies it" (Trexler, 2015). Shelby's novel fits firmly within that tradition.

In sum, *South Pole Station* offers a layered, witty, and compassionate vision of *deep adaptation*. It reminds readers that preserving nature and facing climate change requires not just data and policy, but emotional honesty, cultural openness, and creative reimagining of what it means to coexist on a fragile planet.

Storytelling Through Collapse: Reimagining Futures with Deep Adaptation.

Deep Adaptation Narratives *in Cli-Fi* underscores the vital role of storytelling in confronting the ecological crises of our time. These narratives do more than depict environmental collapse—they offer pathways for reimagining human relationships with nature, community, and self. Through themes of transformation, resilience, cooperation, justice, and speculative hope, *Deep Adaptation* literature challenges readers to envision lives beyond extraction and consumption. As climate realities intensify, such stories become essential tools for emotional reckoning, moral reflection, and collective imagining—helping us not only to survive the wreckage but to write new beginnings from within it.

CHAPTER 8

Literary Reflections Through the Lens of *Collapsology*

In recent decades, increasing environmental instability, resource scarcity, and social fragmentation have intensified public and scholarly discourse on the potential collapse of industrial civilization. One emerging field that addresses these concerns is *collapsology*—a transdisciplinary approach to studying the multifaceted risks associated with societal collapse. The term gained prominence through the work of Pablo Servigne and Raphaël Stevens in their seminal book *How Everything Can Collapse* (2015), which laid the foundational framework for understanding collapse not as a sudden event but as a process unfolding through environmental, economic, and socio-political systems. *Collapsology* is not merely an academic exercise; it is a reflective and strategic response to the pressing crises of the 21st century.

Collapsology refers to the systematic study of the potential collapse of industrial and globalized societies due to interconnected crises, particularly those involving climate change, biodiversity loss, resource depletion, and natural disasters. It operates at the intersection of science, ethics, and futurology, aiming not only to analyse collapse scenarios but also to identify mitigation strategies (Servigne & Stevens, 2015). Unlike traditional disciplines, *collapsology* is intentionally transdisciplinary, drawing knowledge and methodology from ecology, economics, anthropology, sociology, political science, psychology, history, and even the arts. Its holistic nature distinguishes *collapsology* from other forms of environmental or socio-political analysis.

Collapsology employs a rich interdisciplinary approach that integrates insights from multiple domains such as biophysics, demography, law, agriculture, health, and art. This breadth allows for a nuanced understanding of systemic risk and complex causality in societal systems (Servigne & Stevens, 2015). For example, an agricultural collapse cannot be understood without accounting for climate variability, trade policy, soil biology, and social governance.

Unlike historical studies of isolated or ancient civilizations, *collapsology* focuses specifically on the vulnerabilities of contemporary societies. This includes systems heavily dependent on fossil fuels, global supply chains, technological infrastructures, and centralized governance models. As modern civilization grows more interconnected, collapsologists argue that it becomes simultaneously more fragile (Scranton, 2015).

Collapsologists aim to develop foresight tools for identifying early warning signs of collapse and propose adaptive strategies. These may involve community resilience building, decentralized economies, ecological restoration, and shifts in worldview that foster sustainability and cooperation (Diamond, 2005; Morton, 2016).

Collapsology does not merely theorize the end of the world; rather, it encourages strategic foresight and social innovation. By examining historical precedents and current trajectories, *collapsology* urges societies to consider alternative futures rooted in resilience, localization, and ecological harmony. As the field evolves, future research might explore new governance models, degrowth economies, and adaptive infrastructures that respond effectively to systemic shocks.

Moreover, *collapsology* invites a cultural transformation—moving away from anthropocentric and techno-optimist narratives towards more integrative worldviews. Such transformation involves rethinking our relationship with the planet, our definitions of progress, and our responsibilities to future generations.

This essay provides a critical examination of four major works in climate fiction and *collapsology-How everything Can Collapse* (2015), Collapse (2005), Learning to Die in the *Anthropocene* (2015), Dark Ecology (2016)- that explore the concept of societal collapse in the face of climate change and ecological degradation. By analysing these texts, this essay aims to shed light on the ways in which *collapsology* informs our understanding of the risks and consequences of collapse, as well as the possibilities for resilience and transformation.

1.Pablo Servigne and Raphaël Stevens – *How Everything Can Collapse.*

How Everything Can Collapse by Pablo Servigne and Raphaël Stevens (2015) is a pioneering work that introduces the concept of *collapsology*, a transdisciplinary field aimed at studying the potential collapse of industrial and globalized societies. The authors argue that the convergence of ecological degradation, climate change, energy depletion, and economic instability renders societal collapse not just possible but increasingly probable. Unlike traditional academic approaches, this book adopts a holistic and anticipatory perspective, aiming to bridge scientific knowledge with civic responsibility (Servigne & Stevens, 2015).

A key strength of the work lies in its interdisciplinary foundation. Servigne and Stevens draw from a vast range of fields—including ecology, sociology, systems theory, economics, and history—to analyse collapse as a systemic and nonlinear phenomenon. They reject deterministic or apocalyptic views, instead presenting collapse as a process marked by complexity, uncertainty, and multiple tipping points. Their approach challenges the myth of perpetual progress and emphasizes the fragility of modern civilization's dependence on finite resources and global interconnectivity (Servigne & Stevens, 2015, pp. 28–34).

One of the central themes of the book is the concept of "collapsosophy," a way of thinking that combines emotional, ethical, and rational engagement with the prospect of collapse. Rather than provoking despair or denial, collapsosophy encourages a transformative awareness and collective resilience. The authors stress the importance of preparing mentally, socially, and ecologically for discontinuities that may render current institutions, infrastructures, and lifestyles obsolete (Servigne & Stevens, 2015, pp. 102–115).

Another compelling aspect is the authors' effort to humanize the science of collapse. They argue that emotional responses such as grief, anxiety, or even hope are not obstacles but essential components of facing the future. This psycho-social perspective, often overlooked in scientific literature, gives the book a unique emotional depth and makes it accessible to a broader readership, including activists, educators, and citizens concerned with ecological justice.

However, critics have pointed out that the book lacks empirical case studies and tends to generalize collapse scenarios without delving deeply into geopolitical variations. Moreover, its tone—though not alarmist—sometimes leans toward speculative futurism. Nonetheless, its value lies in framing collapse as a legitimate object of inquiry and action rather than a taboo or fringe concern.

In conclusion, *How Everything Can Collapse* is a foundational text in the study of *collapsology*. Servigne and Stevens provide not only a theoretical lens but also a moral and civic call to action. Their work compels readers to confront uncomfortable truths and consider alternative ways of living and organizing society that are rooted in resilience, care, and ecological awareness. It is a critical starting point for anyone seeking to understand the systemic risks of our time and the urgency of preparing for an uncertain future.

2. Jared Diamond – *Collapse: How Societies Choose to Fail or Succeed.*

Jared Diamond's *Collapse: How Societies Choose to Fail or Succeed* (2005) presents a historical and comparative investigation into the reasons behind societal collapses across civilizations. Although Diamond does not explicitly use the term "*collapsology*," his work significantly contributes to the discourse by identifying environmental degradation and poor decision-making as central drivers of collapse. Through a cross-cultural and chronological exploration—ranging from the ancient Maya and Ancestral Puebloans to the Norse Greenlanders and modern-day Rwanda—Diamond posits that the fate of societies hinges not only on ecological conditions but on how societies respond to those challenges (Diamond, 2005).

A hallmark of Diamond's approach is his five-point framework that includes environmental damage, climate change, hostile neighbours, decreased support from friendly trade partners, and the society's response to these pressures. One of the book's greatest strengths lies in its ability to synthesize data from multiple disciplines—geography, anthropology, and history—while presenting narratives that are compelling and accessible to a general audience. He argues that societies that recognize environmental stress and adopt adaptive strategies can avert collapse, while those clinging to destructive practices are more likely to fail.

Diamond's analysis is particularly resonant today, as the global community grapples with climate change, resource scarcity, and growing inequality. His focus on the importance of societal choices—whether to embrace sustainability or persist in exploitative behaviour—offers a prescriptive message that aligns with collapsological thinking. However, Diamond has been criticized for potentially oversimplifying complex political and cultural dynamics and for placing too much weight on environmental determinism (McAnany & Yoffee, 2010).

Nonetheless, Collapse remains an influential work in understanding how environmental and societal factors interact. It offers vital insights for policymakers, educators, and environmentalists alike by framing collapse not as an inevitable outcome, but because of human agency and value-driven decision-making. Diamond's exploration helps bridge historical experience with contemporary ecological concerns, making it a cornerstone text in any collapsological reading list.

3. Roy Scranton – Learning to Die in the *Anthropocene.*

Roy Scranton's ***Learning to Die in the Anthropocene****: Reflections on the End of a Civilization* (2015) blends memoir, philosophy, and environmental science to argue that human civilization must come to terms with the inevitability of ecological collapse. Unlike other works that advocate for technological solutions or societal transformation, Scranton adopts a more existential and stoic stance. He suggests that learning to accept the end of industrial civilization—rather than denying it—is a necessary psychological and cultural step toward living ethically in the *Anthropocene* (Scranton, 2015).

Drawing on his experience as a U.S. soldier in Iraq and his academic background in philosophy and literature, Scranton offers a deeply personal and reflective meditation on death, impermanence, and the hubris of human exceptionalism. He asserts that the modern world is built on fossil fuels and unsustainable consumption, and that no amount of innovation can fully avert the planetary consequences of this legacy. In this context, he calls for a new kind of humanism, one grounded in humility, intellectual courage, and cultural renewal.

One of the most distinctive aspects of Scranton's work is its engagement with literature and the humanities. He argues that stories, myths, and philosophies from dying civilizations carry essential lessons about resilience, acceptance, and meaning-making. Thus, the book does not offer policy proposals or technical blueprints; instead, it offers an ethical and emotional framework for facing collapse. This introspective approach has both inspired and unsettled readers, as it challenges the dominant narratives of progress and control.

While some critics have found Scranton's tone overly fatalistic or lacking in concrete solutions, others praise the book for its raw honesty and philosophical depth. His call to "learn to die" is not a nihilistic surrender but a profound critique of the denialism and false optimism often found in mainstream environmental discourse.

In summary, Learning to Die in the Anthropocene contributes to collapsology by reframing collapse as a moral and existential condition, not merely a material or structural failure. Scranton's work invites us to confront our collective mortality and to find meaning, not in avoiding collapse, but in navigating it with dignity and consciousness.

4. Timothy Morton – *Dark Ecology: For a Logic of Future Coexistence.*

Timothy Morton's Dark Ecology: For a Logic of Future Coexistence (2016) offers a provocative philosophical rethinking of ecological awareness in the Anthropocene. Rather than proposing strategies for sustainability or outlining causes of collapse, Morton delves into the ontological and aesthetic dimensions of ecological thought. He argues that true ecological thinking must move beyond the binary of nature versus culture and confront the unsettling intimacy between humans and the systems they disrupt (Morton, 2016).

Morton introduces the concept of "dark ecology" to describe an ecological philosophy that embraces contradiction, ambiguity, and entanglement. In contrast to "bright green" environmentalism, which often idealizes nature and promotes harmony, dark ecology accepts the discomforting reality that humans are inextricably linked to the very ecological crises they try to solve. For Morton, recognizing this entanglement is essential for developing a more realistic and ethical relationship with the non-human world.

Stylistically, Morton's writing is dense, poetic, and infused with references to literature, music, and continental philosophy. He draws on thinkers such as Heidegger and Derrida, making the text challenging but rewarding for readers with an interest in theory. His ideas challenge traditional ecological thought, proposing that rather than distancing ourselves from toxic environments, we must learn to coexist with them in new ways.

One of Morton's most influential contributions is his critique of "Nature" as a reified and romanticized concept. He contends that the belief in an idealized, external nature perpetuates ecological alienation. Instead, he calls for a flattening of ontological hierarchies—what he calls "ecological thought"—where humans, animals, plastics, and pollution are all part of the same mesh of existence.

Although not a collapsologist in the conventional sense, Morton's work is crucial for understanding the philosophical and affective dimensions of collapse. He encourages readers to abandon the illusion of control and embrace a more complex, interconnected vision of existence. Critics may find the text abstract or inaccessible, but its intellectual ambition and originality offer new ways of thinking about ecology, collapse, and the future.

In essence, Dark Ecology enriches the field of *collapsology* by expanding it into the realm of ethics, aesthetics, and ontology. Morton's insights underscore the importance of reimagining coexistence in a world where the boundaries between human and non-human are increasingly blurred.

Facing the Unthinkable: Collapsology as a Compass for Uncertain Futures.

In sum, *collapsology* provides a critical framework for understanding the systemic vulnerabilities of modern industrial civilization. Drawing from diverse disciplines and cultural traditions, it offers both analytical depth and ethical provocation. The works of Servigne and Stevens, Diamond, Scranton, and Morton illuminate different dimensions of collapse—from empirical case studies to philosophical inquiry—demonstrating that the study of collapse is not only relevant but essential to navigating the uncertainties of the 21st century. By engaging with *collapsology*, societies can cultivate a more reflective, resilient, and ethically grounded approach to the future.

CHAPTER 9

Reclaiming Harmony Through *Eutierria* in Modern Literature

In an age marked by ecological crisis, emotional detachment from nature, and the alienation of humans from the ecosystems that sustain them, Australian philosopher Glenn Albrecht offers a refreshing lexicon to redefine our relationship with the Earth. One of his most evocative contributions is *Eutierria,* coined in his seminal work *Earth Emotions: New Words for a New World* (2019). Derived from the Greek roots eu (good), *terra* (earth), and -ia (a state or condition), *Eutierria* describes a state of deep ecological harmony—a profound emotional and existential connection between humans and the Earth, where distinctions between the self and the environment dissolve in mutual belonging.

This concept emerges from Albrecht's broader vision of the Symbiocene, a proposed future era defined by cooperative coexistence and symbiotic relationships between humans and nonhuman nature. *Eutierria*, then, is not just a feeling but a way of being—a counterforce to the alienation and disconnection characteristic of the Anthropocene. It echoes the emotional kinship of biophilia (love of life), topophilia (love of place), and soliphilia (love of solidarity and responsibility for place), offering a hopeful narrative amidst environmental degradation.

In modern literature, particularly within the frameworks of ecofeminism and posthumanism, the concept of *Eutierria* finds powerful resonance. Ecofeminist literature often critiques the patriarchal structures that have historically justified the exploitation of both women and nature, advocating instead for relational, nurturing, and non-hierarchical modes of interaction. *Eutierria* embodies this spirit, positioning humans not as dominators but as cohabitants of a shared biosphere.

Likewise, posthumanist literature challenges the anthropocentric assumptions that place humans at the center of all value systems. In posthumanist narratives, the boundaries between the human and nonhuman blur, inviting more inclusive, interconnected, and ecological understandings of identity and agency. Characters experience transformation through interactions with animals, landscapes, and ecosystems—experiences that reflect a euteric awareness of oneness with the Earth.

Thus, *Eutierria* is more than a neologism—it is a radical reorientation of thought, emotion, and narrative. As literature increasingly becomes a space to explore alternative ecological futures, Albrecht's concept provides a powerful emotional and philosophical foundation for envisioning a world where humans live not above nature but within it, as part of a larger, vibrant, interconnected whole.

In the realm of modern literature, *eutierria* functions as both a theme and a transformative potential. It aligns with ecofeminist critiques of domination and posthumanist visions of entanglement, offering a shared language of connection in a fragmented world. Whether through Indigenous wisdom, speculative fiction, or environmental philosophy, *eutierria* restores the emotional intimacy between humans and Earth—an intimacy necessary for collective healing.

In naming this emotion, Glenn Albrecht does more than introduce a new term; he revives an ancient truth: that we are not apart from nature but a part of its pulse and poetry. And in the hands of modern writers, that truth becomes a story worth telling—and a world worth saving.

1. *Earth Emotions: New* Words for a New World by Glenn Albrecht.

In Earth Emotions: *New Words for a New World* (2019), philosopher and environmental scholar Glenn Albrecht presents a compelling argument: To address the ecological crises of the Anthropocene, we must also address the emotional and linguistic crises that underlie it. The book offers not only a critique of environmental degradation but a revolutionary new language to express our evolving relationship with the Earth. Albrecht introduces a series of neologisms—most notably, *eutierria*—to capture the emotional spectrum of our ecological experiences.

At the core of Glenn Albrecht's Earth Emotions is the groundbreaking concept of psychoterratic states—a term he uses to capture the full spectrum of emotions humans experience concerning the Earth. These states encompass both negative and positive emotional responses to environmental change. Negative psychoterratic states, such as solastalgia—the distress felt when one's home environment is degraded—and tierratrauma, the psychological impact of ecological disasters, reflect the deep sense of loss and anxiety induced by environmental collapse. In contrast, Albrecht introduces positive psychoterratic states to envision emotional experiences rooted in ecological harmony and hope. These include *eutierria*, the profound sense of oneness with nature; soliphilia, the love for the interconnected Earth community; sumbiophilia, the joy of coexisting with all life forms; and endemophilia, a deep affection for one's native place and its unique ecology. Albrecht contends that the current emotional vocabulary is inadequate to express our complex psychological reactions to planetary change. By naming these feelings, he argues, we can begin to understand, articulate, and transform them into meaningful environmental action and awareness.

Albrecht proposes the Symbiocene as a replacement for the Anthropocene. While the latter is marked by human dominance and ecological degradation, the Symbiocene is envisioned as a new epoch of symbiosis, cooperation, and reconnection between humans and nature. It is within this era that *eutierria* can flourish—a world where humans live not in conquest over nature but in partnership with it.

This vision is not merely ecological but emotional, ethical, and cultural. The Symbiocene represents a future grounded in re-integration, a counterpoint to the alienation that characterizes modern existence.

Albrecht describes *eutierria* as a *positive psychoterratic* state in which individuals feel a *deep sense of harmony, belonging,* and unity with the Earth. It evokes the experience of standing in a forest, swimming in a river, or watching the stars—moments where the boundaries between self and environment dissolve.

Eutierria challenges anthropocentric worldviews by shifting emotional centrality from self-assertion to ecological humility. It reflects a posthumanist and ecofeminist orientation, acknowledging that human well-being is inextricably linked to the health of the planet. This concept resonates with spiritual traditions, Indigenous worldviews, and poetic expressions of "oneness" with nature.

Albrecht's work has significant implications for modern literature, particularly within eco-critical, ecofeminist, and posthumanist frameworks. Writers like Robin Wall Kimmerer, Richard Powers, and Margaret Atwood articulate similar emotions in their narratives—whether through ritualistic gratitude in *Braiding Sweetgrass,* tree consciousness in The Overstory, or speculative futures in *Madd Addam.*

By offering a new lexicon, Albrecht equips readers and writers alike to name and narrate their environmental emotions. In doing so, he fosters a cultural shift—from grief to hope, from alienation to belonging.

A key strength of Glenn Albrecht's Earth Emotions lies in its innovative conceptual framework, which introduces a new emotional vocabulary for engaging with the planet's ecological reality. The book is both accessible and profound, appealing to scholars, activists, and general readers alike, making complex ideas understandable without diluting their depth. While some critics may view Albrecht's vision of the Symbiocene as utopian or overly idealistic, he addresses this critique directly, asserting that utopian thinking is not only valid but necessary for imagining and constructing a more sustainable and harmonious future. His work calls for a paradigm shift in how we perceive and relate to the Earth, grounded in emotional awareness and linguistic innovation. By weaving together insights from philosophy, ecology, psychology, and linguistics, Earth Emotions establishes itself as a foundational text in environmental humanities, offering a compelling roadmap for both intellectual inquiry and emotional resilience in the face of ecological crisis.

Glenn Albrecht's Earth Emotions is a seminal contribution to how we think, feel, and speak about the Earth. By defining *eutierria* and other psychoterratic states, Albrecht does more than enrich our vocabulary—he reorients our moral compass. The book offers a path not just toward environmental awareness but also toward emotional reintegration, linguistic innovation, and cultural renewal.

In an era marked by *ecological grief* and disconnection, Earth Emotions reminds us that healing begins with feeling and that new worlds become possible when we dare to name them.

2. *Braiding Sweetgrass* by *Robin Wall Kimmerer*.

Robin Wall Kimmerer's *Braiding Sweetgrass* is a remarkable work that weaves together Indigenous wisdom, scientific knowledge, and the teachings of plants into a tapestry of ecological consciousness and reciprocity. As a member of the Citizen Potawatomi Nation and a distinguished professor of environmental biology, Kimmerer brings a unique perspective that bridges the gap between Western science and Indigenous ways of knowing.

The book is structured as a collection of essays, each exploring themes of gratitude, interconnectedness, and the sacred relationship between humans and the natural world. Kimmerer introduces readers to the concept of "honorable harvesting," emphasizing the importance of taking only what is needed and giving back to the Earth in return. This principle challenges the exploitative practices of modern consumerism and advocates for a more sustainable and respectful approach to resource use.

One of the most compelling aspects of the book is its poetic and reflective tone. Kimmerer seamlessly blends scientific explanations with personal anecdotes and Indigenous storytelling, creating a narrative that is both informative and deeply moving. For instance, she recounts the story of Skywoman, an Indigenous creation myth, to illustrate the values of generosity and care for the Earth. She also delves into the ecological significance of sweetgrass, a plant revered for its healing properties and spiritual symbolism.

Kimmerer's writing is imbued with a sense of hope and possibility. She envisions a future where diverse knowledge systems coexist and complement each other, fostering a deeper understanding of our place in the web of life. Her call for a "covenant of reciprocity" invites readers to reimagine their relationship with the natural world, not as dominators but as partners and caretakers.

The book has been widely praised for its ability to inspire ecological awareness and action. It resonates with readers who are seeking a more meaningful connection with nature and a path toward environmental sustainability. *Braiding Sweetgrass* is not just a book; it is a guide to living in harmony with the Earth and embracing the wisdom of plants.

The themes of reciprocity and interconnectedness in *Braiding Sweetgrass* resonate deeply with your work on ecofeminism and envisioning sustainable futures. Kimmerer's concept of reciprocity, which emphasizes a give-and-take relationship with the Earth, parallels ecofeminist critiques of exploitative systems that disregard the interconnectedness of all life. Her work invites a rethinking of human relationships with nature—not as hierarchical but as cooperative and reciprocal.

Ecofeminism often highlights the parallels between the domination of women and the exploitation of nature under patriarchal and capitalist systems. Kimmerer's vision of a "covenant of reciprocity" challenges these dynamics by advocating for mutual respect and care, which echoes the ecofeminist call for a more just and equitable relationship with the Earth. Similarly, her emphasis on gratitude and interconnectedness aligns with the ecofeminist focus on valuing nonhuman entities as active participants in the web of life.

In your exploration of ecological consciousness and sustainable communities, Kimmerer's narrative provides a compelling framework for integrating Indigenous wisdom with contemporary ecological practices. Her work underscores the importance of diverse knowledge systems—a theme that likely resonates with your interdisciplinary approach, particularly your engagement with posthumanist and ecofeminist perspectives.

Moreover, *Braiding Sweetgrass* offers fertile ground for imagining alternative futures where respect, care, and cooperation replace exploitation and dominance. It complements your interest in narratives that critique modernity while offering hopeful visions for resilience and sustainability.

Kimmerer's insights offer profound opportunities to enrich both pedagogical and Tanalytical approaches in ecofeminist literature, aligning seamlessly with your commitment to fostering ecological consciousness and sustainable futures.

Kimmerer emphasizes the importance of gratitude, reciprocity, and active engagement with the natural world. Incorporating these principles into ecofeminist pedagogy could foster experiential learning methods, such as fieldwork or reflective practices, that allow students to connect intellectually, emotionally, and physically with the environment. Her blending of Indigenous wisdom with scientific knowledge encourages the inclusion of diverse epistemologies in teaching. This could involve exploring Indigenous creation myths, like Skywoman, alongside ecofeminist texts to highlight the intersections between narrative, ecology, and culture. By centering a relational ethic that values nonhuman agency, Kimmerer's work could inspire discussions on the moral responsibilities of humans toward the environment, enriching the ethical dimensions of ecofeminist curricula.

The concept of reciprocity could serve as a critical framework to analyze ecofeminist texts. For instance, how do narratives in ecofeminism reflect or challenge the idea of mutual exchange between humans and nature? Kimmerer's portrayal of interconnectedness aligns with your focus on posthumanist themes. Her work could inform analyses that deconstruct anthropocentric boundaries and explore interdependencies in ecofeminist literature. Her critique of consumerist and extractive practices mirrors ecofeminist critiques of capitalist systems. This could provide a deeper understanding of how literary texts critique modernity and propose alternative, sustainable futures.

Braiding Sweetgrass is not merely a book—it is an invitation to live differently. With its blend of poetic insight, ecological wisdom, and spiritual grounding, *Robin Wall Kimmerer*'s work challenges, heals, and inspires. In an age of climate crisis and ecological alienation, this book reminds us that the Earth is not a commodity but a gift—and that the most radical thing we can do is to remember how to give back.

3. *The Nature Fix: Why Nature Makes Us Happier, Healthier, and More Creative* by Florence Williams.

Florence Williams's *The Nature Fix* is an engaging and meticulously researched exploration into the transformative effects of nature on the human mind and body. Drawing on a wealth of scientific studies, global field trips, and interviews with leading experts, Williams presents a compelling case for the healing power of the natural world.

Williams takes readers on a journey across the globe—from forest bathing in Japan to urban green spaces in Finland—to investigate how and why nature is crucial to human well-being. Each chapter is rich with anecdotes, science-based evidence, and personal reflections, making complex research both accessible and relatable.

One of the central themes of the book is how nature reduces stress, anxiety, and depression. Williams discusses how exposure to natural environments helps recalibrate our nervous systems, lowers cortisol levels, and promotes physiological recovery. The book builds a strong argument for incorporating green spaces in daily life, not just for leisure but as a public health necessity.

The connection between human health and the environment underpins many of the studies Williams explores. This includes references to animal-assisted therapy, horticultural therapy, and biophilic design in architecture—all of which support the idea that our bodies and minds respond positively to natural elements.

In addition to emotional well-being, *The Nature Fix* delves into cognitive benefits. Williams reviews studies that demonstrate how nature enhances concentration, creativity, memory retention, and problem-solving abilities. She highlights the concept of "attention restoration theory", which suggests that natural environments replenish our capacity for focused attention, depleted by overstimulation in urban settings.

Her findings indicate that even brief encounters with nature—such as a walk in a park or viewing greenery from a window—can significantly improve mental clarity and emotional balance.

Beyond the scientific data, Williams emphasizes the aesthetic and emotional connection humans have with nature. This connection to place, landscape, and environment often evokes a sense of wonder, joy, and belonging, which are deeply therapeutic and spiritually enriching.

Florence Williams's *The Nature Fix*: Why Nature Makes Us Happier, Healthier, and More Creative explores the profound impact of nature on mental and physical well-being. The book is praised for its engaging writing style and thorough investigation into the science behind nature's restorative benefits. Williams delves into research from various locations, such as forest trails in Korea and islands in Finland, to uncover how exposure to natural environments enhances mood, creativity, and health.

Readers appreciate her ability to balance scientific rigor with accessible storytelling, making complex topics relatable. The book emphasizes the urgency of reconnecting with nature in our increasingly indoor lifestyles. It's a compelling read for anyone interested in the intersection of environment and human well-being.

At the heart of *The Nature Fix* is the concept of biophilia, a term popularized by biologist E.O. Wilson, which refers to the innate human affinity for nature. Williams explores how this deep-seated connection is not merely poetic but profoundly biological and psychological. She explains that humans evolved in natural environments, and thus, our brains are hardwired to respond positively to natural stimuli—like the rustling of leaves, the smell of pine, or the sound of water.

However, as modern lifestyles increasingly shift indoors and toward digital screens, this bond is weakening. Williams warns that this disconnect comes at a cost, including higher rates of stress, depression, anxiety, and even cognitive fatigue. She stresses the importance of consciously reintroducing nature into our daily routines to reignite our biological need for the wild.

One of the most captivating sections of the book covers Shinrin-yoku, or forest bathing, a Japanese practice that involves leisurely walks through forests while engaging the senses. Williams delves into the science behind forest therapy, citing studies showing that even a couple of hours in a forest can significantly reduce cortisol (the stress hormone), lower blood pressure, and enhance immune function by increasing the activity of natural killer cells.

She interviews researchers and participates in forest bathing herself, making the case that immersing oneself in the sensory experience of the woods—the sounds, smells, sights, and textures—can lead to deep relaxation and mental restoration. This practice, rooted in both tradition and science, becomes a strong example of how nature can serve as a preventive and restorative health strategy.

Williams expands her investigation by exploring how various cultures incorporate nature into daily life. For instance, in Finland, nature is seen as a birthright. The concept of Everyman's Right allows citizens to roam freely through natural spaces. Schools emphasize outdoor learning, and time in the forest is considered vital for children's development. In South Korea, the government actively promotes forest healing centers as part of public healthcare. In Singapore, nature is embedded in urban design, with towering vertical gardens and tree-lined streets. These cross-cultural insights underline the fact that different societies have found diverse, meaningful ways to maintain a connection to the natural world, offering readers practical models to emulate in their own lives.

With urban populations rising, Williams tackles the question: How can we bring nature to the city? She explores innovative approaches to urban greening, such as Green rooftops that provide insulation, reduce urban heat, and offer mini-ecosystems.

Living walls—vertical gardens that improve air quality and aesthetics. Pocket parks and green corridors weave nature into the urban fabric. Williams emphasizes that nature doesn't have to be wild and remote to be effective. Even small green spaces can have measurable psychological benefits, like reduced aggression, better mood, and enhanced community cohesion. Her message is clear: Urban planning must prioritize nature access if we hope to support healthier, happier city dwellers.

One of the most empowering aspects of *The Nature Fix* is how it brings nature within reach for everyone. Williams provides tangible strategies for reconnecting with the natural world, even in concrete-heavy environments. Some of her suggestions include taking daily walks in green areas, even if it's just a tree-lined street or a local park. Listening to natural sounds like birdsong or running water, which can have a calming effect, even when experienced through recordings. Adding plants to indoor spaces improves air quality and mood. Practicing mindful observation during outdoor moments—savoring the colors of leaves, the breeze on your skin, or the shape of a cloud.Planning nature getaways to rejuvenate the mind and body, even if it's just a weekend hike. These tips are grounded in science yet feel approachable, reinforcing the idea that you don't have to escape to the wilderness to benefit from nature—you just have to notice it.

The Nature Fix is not only a report on nature's impact—it's a call to action. Williams challenges readers to reconsider their relationship with the natural world, advocating for policy changes, educational reforms, and personal habits that nurture this essential bond. Her exploration of biophilia, forest therapy, urban greening, and cultural practices illustrates that a healthier, more joyful life is possible—if we let nature back in.

The Nature Fix is more than just a presentation of scientific facts; it is a passionate plea to reconsider our relationship with the natural world. Williams argues that reconnecting with nature isn't a luxury—it's essential for our survival and thriving in a fast-paced, technologically saturated world.

This book is highly recommended for educators, policymakers, mental health professionals, and anyone interested in the intersection of health, psychology, and the environment. By blending journalism, personal narrative, and cutting-edge science, Florence Williams offers a persuasive and inspiring case for returning to nature as a path toward happiness, health, and human potential.

Literary Pathways to Eutierria: Reclaiming Harmony with the Earth.

As environmental degradation and disconnection from nature become defining concerns of our time, modern literature serves as a powerful medium to restore the lost harmony between humans and the Earth. The concept of *Eutierria*—a profound sense of unity with the natural world—finds expression in the works of contemporary writers who challenge human-centered worldviews and reimagine our relationship with the planet. Authors like Glenn Albrecht, *Robin Wall Kimmerer*, Richard Louv, Richard Powers, and Amitav Ghosh explore this interconnectedness through deeply reflective narratives that foster ecological awareness and emotional resonance with nature. These texts not only critique the ecological consequences of modern lifestyles but also offer a vision of hope rooted in empathy, reverence, and coexistence. By evoking *Eutierria*, modern literature invites readers to reconnect with the Earth, recognize the sacredness of all life, and embrace a more sustainable and harmonious way of living.

CHAPTER 10

Climate Refugees: Displacement in the Age of Climate Change

The phenomenon of climate-induced displacement has emerged as one of the most urgent humanitarian and legal challenges of the 21st century. As global awareness of climate change and its devastating consequences continues to rise, so too does the use of the term "*climate refugees*." However, defining this term remains complex due to the multifaceted nature of climate migration, which often involves a range of environmental, economic, and socio-political factors. Climate change—manifesting through rising sea levels, extreme weather events, droughts, floods, and soil salinization—renders certain regions uninhabitable, forcing vulnerable populations to migrate both internally and across borders. Despite their growing numbers, *climate refugees* fall outside the traditional scope of the 1951 Geneva Convention, which narrowly defines a refugee as someone fleeing persecution based on race, religion, nationality, political opinion, or membership in a particular social group. This legal gap has sparked intense debate within international law and human rights circles, leaving climate-displaced individuals in a precarious position with limited protection.

The term "*climate refugees*" is not officially recognized under international law, which complicates their protection. Unlike refugees fleeing persecution or conflict, *climate refugees* often remain within their own countries, making them "internally displaced persons" rather than cross-border migrants. However, the scale of displacement is staggering, with weather-related disasters triggering over 32.6 million internal displacements in 2022 alone.

World literature plays a vital role in humanizing the crisis of climate-induced displacement by transcending the limitations of statistics and policy discourse to foreground the emotional, cultural, and ethical dimensions of the refugee experience. While scientific reports and media coverage provide valuable data on rising sea levels, droughts, and mass migrations, literature enables readers to engage with the lived realities of those affected. Through fiction, poetry, memoirs, and speculative narratives, writers from diverse cultural contexts give *climate refugees* names, stories, and voices, thus transforming them from abstract figures into fully realized human beings. World literature amplifies silenced and marginalized voices—particularly those of indigenous peoples, women, and communities from the Global South—whose experiences are often overlooked in global debates. By portraying localized responses to global phenomena, literature bridges the gap between the global and the personal, fostering a sense of interconnectedness and shared responsibility. Furthermore, it mediates between memory and migration, preserving cultural identities through narratives that reclaim lost homelands. In addition, speculative and climate fiction allow writers to imagine dystopian futures shaped by environmental collapse, serving both as cautionary tales and as tools for reimagining more just, sustainable worlds. Importantly, world literature raises urgent ethical and political questions about accountability, inequality, and environmental justice, encouraging readers to reflect on their roles within systems of power. Above all, literature cultivates empathy, compelling readers to see *climate refugees* not as distant victims but as fellow humans navigating trauma, resilience, and hope. In this way, world literature becomes a powerful medium for global citizenship, advocacy, and ecological awareness, reminding us that the climate crisis is not merely an environmental issue but a profound human and moral challenge.

In the unfolding crisis of climate change, literature has become a powerful medium to articulate the voices of those who face the brunt of ecological devastation. *Climate Refugees: Displacement in the Age of Climate Change* explores how literary narratives portray the trauma of losing one's ancestral land due to rising sea levels, desertification, and extreme weather. These stories are deeply infused with themes of displacement and the emotional weight of nostalgia and longing for lost homes. At the heart of many such works is a critique of environmental injustice, spotlighting how the global South disproportionately bears the consequences of a climate emergency largely driven by industrialized nations of the global North. Literature also draws attention to the intersectional dimensions of this crisis, examining how race, class, and gender compound vulnerability and suffering. Yet, amidst the despair, there are stories of survival, resistance, and resilience—narratives that celebrate indigenous knowledge systems, community solidarity, and the human capacity to adapt. Forced migration, as depicted in literary works, is not just a movement of bodies but of memories, traditions, and identity. Moreover, speculative and dystopian fiction offers provocative visions of future refugee crises sparked by ecological collapse. These imagined worlds serve as urgent warnings and wake-up calls to current generations, urging reflection and action before fiction becomes reality.

Literature serves as a powerful medium for cultivating empathy and raising awareness, allowing readers to grasp the human dimensions behind the abstract statistics of climate change and displacement. Many writers across the globe have harnessed the art of storytelling to illuminate the pressing issue of *climate refugees*, offering narratives that intertwine environmental catastrophe with personal and collective trauma. This article explores how select literary works engage with themes of climate-induced migration and ecological crisis. Notable among them is *Amitav Ghosh's Gun Island (India),* which intertwines myth, migration, and climate change in a narrative that spans continents. *Helon Habila's Oil on Water (Nigeria)* delves into the environmental devastation of the Niger Delta and its impact on local communities. Octavia *Butler's Parable of the Sower (USA)* offers a dystopian vision of a climate-ravaged future where social collapse forces migration and spiritual rebirth. *Mohsin Hamid's Exit West (Pakistan/global),* though not strictly climate fiction, allegorically explores displacement and the fluidity of borders in times of crisis. Finally, *the Refugee Tales series (UK)* presents a powerful collection of narratives based on real-life experiences of displaced people, echoing the emotional and psychological toll of forced migration. Through these works, literature transcends geographical boundaries to present a shared global concern, urging readers to confront the realities of climate-induced displacement and rethink our collective responsibility.

1. *Gun Island* (India) by Amitav Ghosh: A Lyrical Confluence of Myth, Migration, and Climate Crisis

Amitav Ghosh's *Gun Island* (2019) is a masterful blend of mythology, migration, and environmental consciousness, positioning itself as one of the most significant literary responses to the climate crisis in contemporary world literature. Ghosh, known for his interest in the entanglements of history, colonialism, and ecology—as seen in earlier works like *The Hungry Tide* and the Ibis Trilogy—takes a bold leap in *Gun Island* by infusing speculative and folkloric elements into the realm of climate fiction, or cli-fi." In doing so, he challenges the conventions of realist fiction, insisting that the climate crisis demands new narrative forms and imaginative interventions.

At the heart of Gun Island lies the journey of Deen Datta, a middle-aged Bengali rare book dealer living in Brooklyn, whose ordinary life is disrupted by an encounter with a half-forgotten Bengali legend—the tale of the merchant Chand Sadagar and the snake goddess Manasa Devi. What begins as a curious scholarly inquiry soon spirals into a globe-spanning journey from the mystical Sundarbans of Bengal to the canals of Venice, reflecting a shift not only in geographical space but also in epistemological understanding. As the narrative progresses, Deen's skepticism is challenged by a series of surreal events, leading him to embrace a world where myth, history, and contemporary crises intersect. Ghosh deftly weaves together the personal and the planetary, showing how ancient stories hold relevance in our modern world, which is plagued by ecological devastation and forced displacement.

Thematically, Gun Island foregrounds issues of migration, displacement, belonging, and resilience, not only through human experiences but also through the lens of non-human agency. Characters such as Cinta, an aging yet sharp Venetian scholar, and Tipu, a tech-savvy, undocumented Bengali migrant, bring diversity to the narrative and embody the global flows of people, knowledge, and cultures. The novel repeatedly asks: What happens when the very landscapes people call home become hostile to life due to environmental collapse?

Gun Island is profoundly ecocritical, engaging with environmental degradation, loss of biodiversity, and the trauma of climate-induced displacement. Ghosh presents *climate refugees* not just as casualties of natural disasters but as participants in a much larger, interconnected ecological drama. The narrative draws attention to the migrant ecologies of both humans and animals, pointing to how climate change disrupts established life patterns across species. In this sense, the novel resists anthropocentrism and advocates for a multi-species environmental consciousness, urging readers to reconsider the boundaries between the human and the non-human.

Ghosh critiques the asymmetrical burden of climate change, where impoverished and marginalized communities—often in the Global South—suffer disproportionately despite contributing the least to the environmental crisis. From rising sea levels in the Sundarbans to heat waves and forest fires in California, *Gun Island* provides a mosaic of environmental emergencies, illustrating how the impacts of climate change transcend borders and class divides.

Ghosh's deliberate embrace of the surreal and the magical is both a stylistic choice and a political gesture. By invoking uncanny events—such as venomous snakes appearing in unlikely places or Deen's repeated brushes with inexplicable phenomena—Ghosh unsettles readers and refuses the linear logic of Western rationalism. He argues, both implicitly and explicitly, that realism as a literary mode is inadequate to grasp the complexity and urgency of the climate crisis.

The novel thus blurs the boundaries between myth and science, past and present, and East and West. The myth of Manasa Devi becomes an allegory for environmental revenge and resilience, a symbolic framework through which

Ghosh explores the spiritual and emotional dimensions of climate trauma. In choosing to "believe" rather than "explain," the novel invites a rethinking of knowledge systems and challenges the supremacy of empirical science in environmental discourse.

Set against the backdrop of Venice's migrant crisis, *Gun Island* takes a sharp look at the ethical tensions surrounding borders, asylum, and global inequality. Ghosh uses the metaphor of water—rising tides, floods, deluges—to signify both natural upheaval and human movement, collapsing the distinctions between ecological and political crises. The city of Venice, itself threatened by climate change, becomes a powerful symbol of a world struggling to hold back the tide of environmental catastrophe and humanitarian displacement.

Through its characters and settings, *Gun Island* calls for transnational solidarity, cultural dialogue, and ethical responsibility. It champions the forging of new alliances—between the privileged and the dispossessed, between humans and nature—as the only viable response to an increasingly fractured world. The novel insists that climate change is not just an environmental issue but a profound moral and existential challenge, one that requires not only technological solutions but a transformation of consciousness.

In Gun Island, Amitav Ghosh offers more than a narrative; he offers a framework for understanding the intertwined fate of climate, culture, and displacement. The novel is at once a lament and a call to action, a reminder that the stories we tell—and believe—about our world shape the possibilities for its survival. Through his innovative blend of magical realism, mythology, and ecological critique, Ghosh redefines the role of the novelist in the Anthropocene. He does not simply tell us what is happening; he helps us feel its urgency, imagine alternatives, and empathize with those most affected. *Gun Island* stands as a literary testament to the power of fiction to inspire ecological awareness and transnational empathy in the face of a planetary crisis.

2. *Oil on Water* (Nigeria) by Helon Habila: A Stark Portrait of Oil Politics, Environmental Devastation, and Postcolonial Struggles in the Niger Delta.

Helon Habila's Oil on Water (2010) is a compelling and haunting narrative that uncovers the environmental and human tragedies resulting from oil extraction in Nigeria's Niger Delta. In this beautifully crafted yet grim novel, Habila paints a vivid picture of a region ravaged by pollution, violence, and displacement, foregrounding the often-neglected human and ecological cost of global oil consumption. With its fluid narrative and journalistic undertone, the novel transcends fiction to become a powerful political and environmental commentary rooted in lived experiences.

Set in the swamps and creeks of the Niger Delta, Oil on Water follows two journalists—Rufus, a young and idealistic narrator, and Zaq, a disillusioned veteran—on a journey to investigate and report the kidnapping of Isabel Floode, a British oil executive's wife. Their pursuit of truth becomes an odyssey through a landscape scarred by oil spills, military occupation, and ecological ruin. The journey motif allows Habila to explore the multiple layers of violence—environmental, physical, and ideological—that define the region.

Through its episodic structure and fragmented memories, the novel presents a layered narrative that reflects the fractured lives of the Niger Delta's inhabitants. Each encounter with displaced villagers, guerrilla fighters, and oil company representatives adds nuance to the complex web of blame, resistance, and survival. Habila avoids didacticism, allowing the setting itself—a toxic, decaying ecosystem—to speak as eloquently as his characters.

At its core, Oil on Water is an ecocritical work that viscerally portrays environmental degradation as both a literal and metaphorical condition. The Niger Delta is depicted as a dystopian wasteland, where oil slicks coat the river's surface, gas flares light up the night sky, and the once-thriving aquatic life has disappeared. The land is blackened, and the air is poisoned. Fishing and farming, the main livelihoods of the people, have been obliterated, replaced by a desperate struggle for survival.

Habila draws attention to environmental racism, wherein multinational corporations—primarily Western oil giants—collaborate with corrupt local governments to exploit the Delta's resources, leaving behind destruction and despair. This dynamic reflects a broader neo-colonial structure in which the wealth extracted from African soil benefits foreign powers while locals inherit the waste, disease, and dislocation. Communities are rendered *climate refugees* in their land, forced to abandon their ancestral homes due to pollution, violence, and loss of livelihood.

The novel offers a searing critique of postcolonial governance and the collusion between state forces and multinational corporations. Even though Nigeria is politically independent, its economic sovereignty is undermined by its dependence on oil exports and the influence of foreign capital. The government's use of military force to protect oil interests—often at the cost of its citizens—exemplifies the authoritarian grip of petro-capitalism.

Habila introduces readers to the militant resistance movements that have emerged in response—groups often labeled as terrorists or criminals but portrayed in the novel with complexity and moral ambiguity. These groups reflect the frustration and helplessness of communities caught between environmental collapse and political abandonment.

Characters like Professor, a former teacher turned guerrilla leader, symbolize the transformation of intellectual resistance into armed struggle in the face of systemic injustice.

From a literary standpoint, Habila's prose is lyrical yet restrained, blending poetic imagery with journalistic precision. His choice of water as a recurring motif is especially powerful: water is life-giving but also corrupted in the novel, symbolizing both the lifeblood of the land and its contamination. The river, once a site of cultural richness, has become a corridor of death and decay.

The title Oil on Water itself is metaphorically rich—suggesting a superficial gloss that conceals deeper damage, an unnatural mixture, and a tension between what is seen and what lies beneath. Habila's use of metafictional elements—such as the role of the journalists and the challenges of narrating "truth" in a fractured world—adds depth to the narrative, raising questions about representation, memory, and ethical responsibility.

The novel also explores the eco-feminist dimension, showing how women bear a disproportionate share of the suffering. From displacement to sexual violence, women in the Niger Delta are portrayed as silent casualties of a war they did not start but are forced to endure. The violated land and the violated female body are intricately linked, echoing the traditional ecofeminist trope of nature and woman as victims of patriarchal, capitalist exploitation.

Ultimately, *Oil on Water* demands a moral reckoning—not only from its characters but also from its readers. It forces us to confront the hidden costs of the modern world's oil dependence and challenges us to reflect on the lives sacrificed for global convenience and economic gain. The novel humanizes those rendered invisible by international politics and market forces, giving voice to the fishermen, villagers, and activists whose suffering is too often ignored.

In doing so, Habila contributes significantly to the emerging canon of climate literature from the Global South, offering a vital African perspective on ecological trauma and environmental justice. His work aligns with the growing awareness of climate colonialism, in which environmental harm is disproportionately inflicted upon former colonies to benefit the global elite.

Helon Habila's Oil on Water is more than a novel; it is a poignant exposé of how environmental destruction intersects with colonial legacies, political corruption, and social despair. It documents a landscape and a people on the brink, yet it also preserves their stories, their resistance, and their humanity. With eloquence and urgency, Habila invites readers to listen—to the cries of polluted rivers, the silence of abandoned villages, and the resilience of those who refuse to disappear.

Oil on Water is a necessary read for anyone concerned with environmental ethics, postcolonial justice, and the human faces of ecological catastrophe. It exemplifies how literature can confront complex global issues while retaining the intimacy of personal stories, proving that fiction can illuminate truths that data alone cannot convey.

3. *Exit West* (Pakistan / Global): Mohsin Hamid: A Tender, Magical Portrait of Refuge, Loss, and Belonging in a Borderless World.

Mohsin Hamid's *Exit West* (2017) is a luminous and timely novel that reimagines the refugee experience through a fusion of magical realism and political realism. Set in a near-contemporary world on the brink of collapse, the novel follows Saeed and Nadia, a young couple whose lives are disrupted by civil war. What begins as a love story soon transforms into a haunting, meditative journey across borders—both literal and psychological.

What makes *Exit West* singular is Hamid's *radical recon-*

figuration of the migrant narrative, replacing barbed wires and asylum camps with magical doors that whisk people from one part of the world to another. These portals, both dreamlike and devastating, challenge the reader to reflect on the fluidity of movement, the right to migration, and the shared human experience of longing, loss, and renewal.

At its core, Exit West is an allegory of forced migration. The city where Saeed and Nadia meet is unnamed—deliberately so—to underscore the universality of war and displacement. As violence erupts and the city falls into chaos, the couple flees, becoming part of a global tide of refugees traveling through mysterious doors to Mykonos, London, and San Francisco.

Hamid avoids the gritty literalism typical of refugee fiction. Instead, he distills the emotional truth of displacement: the trauma of leaving home, the fear of the unknown, the ache of nostalgia, and the disorientation of forging a new identity in a foreign land. These doors become symbolic of escape and access, but also of disruption, alienation, and the surreal quality of modern transience.

Through this narrative strategy, Hamid reimagines migration not as a linear passage from suffering to salvation but as a perpetual transformation marked by uncertainty, fragmentation, and resilience.

One of the most profound aspects of Exit West is its attention to the psychological toll of displacement. Saeed and Nadia, though close at first, begin to drift apart as they traverse new geographies. Their emotional detachment mirrors the identity dislocation that refugees often face, not just from their homeland but from their former selves.

The novel captures the psychology of exile—a slow erosion of cultural anchorage, intimacy, and shared memory. As they move through different societies, both characters wrestle with belonging, assimilation, and the pain of being seen as "other." Saeed, with his spiritual ties to family and tradition, yearns for home, while Nadia, a more adaptive and secular figure, embraces reinvention. Their diverging responses reflect two modes of survival in exile: preservation and transformation.

Exit West thoughtfully explores the intersection of culture, gender, and autonomy, particularly through Nadia's character. Her refusal to wear traditional dress for religious reasons, despite societal expectations, marks her as an early symbol of resistance. As she navigates different cities and identities, Nadia becomes a figure of cosmopolitan adaptability, unbound by traditional structures.

Hamid positions identity as fluid and plural, not fixed to geography or culture. The novel insists on the human particularity of migrants, countering the dehumanizing tendency of global discourse to reduce refugees to statistics or stereotypes. Through Saeed and Nadia, he shows how migration is not only about physical movement but also about remaking the self in shifting cultural landscapes.

The novel's most remarkable literary technique is the use of magical doors—portals that instantly transport characters across continents. These doors are never explained, which lends them a mythic quality, allowing Hamid to sidestep the literal violence of border crossings while emphasizing the emotional cost of exile.

This surreal device enables a condensed, borderless narrative, mirroring how digital technologies and global crises collapse distances in the real world. Simultaneously, it draws attention to the inequities of mobility—those with privilege can move freely, while others face walls, checkpoints, and detention.

The magic, then, does not erase suffering. It exposes the cruel absurdity of borders and imagines a world where movement is a right, not a crime.

Mohsin Hamid, with his transnational background—born in Pakistan, educated in the U.S., and a citizen of the world—writes from a hybrid position. His prose is sparse yet lyrical, contemplative yet charged. He writes in long, flowing sentences that mimic the drift of thought and the flow of migration. His narrative voice is both intimate and detached, inviting empathy while evading sentimentality.

Hamid resists national allegiances and instead writes toward a global citizenship of compassion. He deconstructs the binaries of East and West, local and foreign, home and exile. In doing so, *Exit West* becomes not just a story of two migrants but a philosophical inquiry into what it means to be human in a world of constant flux.

Hamid's novel also acknowledges the role of technology in shaping migration. Characters use cell phones to navigate routes, stay in touch with loved ones, and access information. These small devices become lifelines of identity and continuity.

Yet, technology also underlines the disparities of globalization. The novel depicts a planet where people are more connected than ever yet increasingly divided by fear, xenophobia, and surveillance. Hamid subtly critiques this paradox, urging greater empathy and structural change.

Exit West closes not with a grand revolution but with a quiet meditation on growth, memory, and divergence. Saeed and Nadia, having reached the end of their shared journey, part ways. There is no forced resolution—only acceptance of the transience of relationships and the inevitability of change.

In imagining a world where borders are irrelevant and where human movement is as natural as breathing, Hamid offers a radical and tender vision of what migration could look like—not a crisis, but a new normal shaped by hospitality, humility, and human connection.

Mohsin Hamid's *Exit West* is a masterpiece of compassion and imagination—a border-defying novel that asks us to reexamine what it means to belong, to migrate, to love, and to let go. Through its luminous prose and visionary premise, the novel speaks urgently to our contemporary moment, offering not despair but the possibility of radical empathy.

In a world increasingly fractured by borders and fear, *Exit West* invites us to believe in doors, not walls, and to walk through them together.

4. *Refugee Tales* – A Modern Pilgrimage of Resistance, Witness, and Solidarity

"No one walks alone." This simple yet profound sentiment lies at the heart of the Refugee Tales project—a unique literary and activist initiative that brings together refugees, asylum seekers, ex-detainees, and established British writers in a shared act of storytelling and protest. Inspired by Geoffrey Chaucer's The Canterbury Tales, this ongoing project (since 2015) reimagines the pilgrimage as a collective, transgressive journey of bearing witness, healing, and demanding justice for the voiceless.

Organized by the Gatwick Detainees Welfare Group, Refugee Tales is a defiant cultural response to the UK's indefinite detention policy and wider hostile environment strategies. It transforms anonymous suffering into narrated presence, turning trauma into testimony and story into solidarity.

The project includes Four volumes (as of 2021) of short stories and life narratives.

Collaborations between refugees/detainees and acclaimed authors (e.g., Ali Smith, Kamila Shamsie, Bernardine Evaristo).Annual protest walks across the English countryside, symbolically and literally challenging border enforcement practices and indefinite detention.Through these artistic and embodied practices, Refugee Tales becomes more than just a literary collection—it becomes a form of mobile resistance, a platform for advocacy, and a ritual of remembrance.

One of the most vital functions of Refugee Tales is its counter-hegemonic narrative work. The stories give form and voice to those rendered invisible by policy and public discourse. Refugees and detainees, often reduced to case numbers or tabloid caricatures, are re-humanized through intimate storytelling that reveals their inner lives, histories, and struggles.

Narrating Displacement: World Literature and the Human Face of Climate Refugees

In conclusion, world literature offers a vital lens through which to understand the complex realities of climate-induced displacement, humanizing the experiences of *climate refugees* and amplifying their silenced voices. By weaving together narratives of environmental degradation, forced migration, and resilience, these literary works challenge dominant political discourses and invite readers to confront the ethical and ecological dimensions of global inequality. As rising seas, droughts, and natural disasters continue to reshape the contours of habitability, world literature not only bears witness to the traumas of uprooted lives but also envisions new modes of belonging, solidarity, and planetary responsibility. In doing so, it compels us to recognize that climate change is not only an environmental crisis but also a profound humanitarian and narrative one.

CHAPTER 11

"Climate Justice and World Literature: Exploring the Intersection"

In an era marked by escalating environmental crises, *climate justice* has emerged as a central concern not only in scientific and political discourse but also within the realm of literature. *Climate justice* seeks to address the disproportionate effects of climate change on marginalized communities, indigenous peoples, and the Global South—those who have contributed least to the problem but are most vulnerable to its consequences. While scientific data and policy frameworks provide essential tools for understanding and responding to climate change, world literature offers a uniquely human perspective, engaging readers emotionally, ethically, and imaginatively with the lived realities of environmental injustice.

This article explores the dynamic intersection between *climate justice* and world literature, recognizing literature as a powerful medium that articulates the human dimensions of ecological disruption. Across cultures and continents, literary texts—novels, poetry, drama, and oral traditions—have long served as mirrors and critiques of socio-environmental realities, illuminating the entangled relationships between environmental degradation, colonial histories, economic inequality, and cultural erasure.

From the tidal landscapes of the Sundarbans in Amitav Ghosh's *The Hungry Tide* to the drought-stricken terrains in African fiction, and from indigenous American eco-poetry to speculative climate fiction (cli-fi) in the West, literary works reflect and reframe climate change not merely as an environmental issue, but as a profound ethical and political challenge. These texts humanize the statistics, foreground stories of resistance and resilience, and foster a planetary consciousness that is crucial for achieving environmental and social justice.

By analyzing how world literature engages with themes of displacement, *ecological grief*, *intergenerational* ethics, and resistance, this article argues that literature plays a vital role in shaping the global conversation on *climate justice.* It not only documents injustices but also *imagines alternative futures,* offering a moral compass in a time of planetary crisis.

Climate justice is a multidimensional concept that brings together environmental sustainability and social equity, emphasizing that the impacts of climate change are not just scientific or technical concerns but fundamentally issues of morality, responsibility, and human rights. At its core, *climate justice* demands the equitable distribution of the benefits and burdens associated with both the causes of climate change and the strategies used to address it—namely, mitigation (reducing emissions) and adaptation (adjusting to effects).

The concept acknowledges that climate change does not affect all people equally. Those who are least responsible for greenhouse gas emissions—such as low-income communities, indigenous populations, and residents of small-island developing states—are often the most vulnerable to its devastating consequences. These include rising sea levels, extreme weather events, food and water insecurity, and forced displacement. This unequal vulnerability is shaped by historical injustices, including colonial exploitation, systemic poverty, and environmental racism.

Climate justice, therefore, insists on recognizing and rectifying these structural inequalities. It frames climate change as a social and ethical issue, requiring that climate policies be inclusive, participatory, and sensitive to the needs and rights of the most affected communities. Rather than adopting a "one-size-fits-all" approach to environmental governance, *climate justice* promotes the principle of "common but differentiated responsibilities and respective capabilities" (CBDR-RC), which holds wealthier, industrialized nations more accountable for their historical and ongoing contributions to environmental degradation.

Importantly, *climate justice* also extends to *intergenerational* equity—the idea that present generations have a moral obligation to protect the planet for future inhabitants—and interspecies justice, recognizing the rights and intrinsic value of nonhuman life. As such, *climate justice* is not merely about reducing emissions but about transforming systems of power, uplifting marginalized voices, and envisioning a more just and sustainable world for all beings.

Literature has always served as a profound medium through which human relationships with nature are expressed, interrogated, and reimagined. In the context of the 21st-century climate crisis, world literature has increasingly engaged with the pressing ethical and political questions surrounding *climate justice.* Across geographies and genres, contemporary writers are using fiction, poetry, drama, and memoir to expose environmental injustices, critique global power structures, and amplify the voices of the marginalized. These narratives humanize the often abstract discourse of climate change, revealing how it intersects with issues of race, class, gender, colonial history, and indigenous sovereignty.

Some notable examples that foreground *climate justice* in world literature include:

1. *The Hungry Tide* by Amitav Ghosh.

Amitav Ghosh's *The Hungry Tide* (2004) is a compelling ecological and cultural narrative set in the Sundarbans, a fragile mangrove ecosystem shared by India and Bangladesh. The novel, celebrated for its lyrical prose and layered storytelling, moves beyond traditional environmental fiction to engage with deeper issues of *climate justice*, postcolonial displacement, gendered agency, and the intersection of science and indigenous knowledge. By blending personal journeys with geopolitical tensions and environmental history, Ghosh crafts a literary terrain that is as shifting and perilous as the tides that shape the Sundarbans themselves.

The setting of the Sundarbans—a UNESCO World Heritage site composed of a dynamic landscape of mudflats, mangroves, and tidal rivers—becomes a living, breathing character in the novel. This unique ecosystem, prone to cyclones, erosion, and floods, serves as a metaphor for ecological vulnerability and human resilience. The novel localizes its narrative within this precarious region while simultaneously referencing broader global issues through characters from Delhi and the United States, such as Kanai and Piya.

The partition of Bengal (1947) and the resettlement crisis following Bangladesh's independence form the political backdrop, particularly through the reference to the Morichjhapi massacre, a historical event where the West Bengal government evicted and killed Bengali refugees who had settled in protected forest lands. This episode becomes a symbol of the conflict between environmental protection and human rights, echoing contemporary debates about conservation versus habitation.

Historically, the Sundarbans have been marked by colonial exploitation, with the British Empire utilizing the region for timber and revenue. Post-independence, the area remained marginalized, a space for dispossessed populations, including refugees and the rural poor. Ghosh captures this layered history, embedding it into the very landscape of the novel and invoking a sense of palimpsestic time—where multiple historical epochs overlap.

The novel also subtly critiques neo-colonial conservation policies, showing how environmental regulations imposed from above can mirror the disregard for local agency seen during colonial rule. This tension is especially evident in the narrative arc concerning the protection of the Irrawaddy dolphins and the subsequent policing of local fishing activities, echoing climate injustice themes.

Ghosh presents the region as a climate hotspot—subject to rising sea levels, cyclonic destruction, and biodiversity loss. However, he resists framing the Sundarbans solely as a victim of climate change. Instead, he draws attention to the indigenous knowledge systems and adaptive strategies employed by the local people. Through characters like Fokir, a poor, illiterate fisherman who guides Piya through the river channels, Ghosh emphasizes experiential and embodied knowledge, challenging Western-scientific hegemony. This environmental consciousness is not didactic but integrated into the narrative fabric, highlighting coexistence, conflict, and negotiation between humans and nature.

The novel critiques nationalist cartographies, particularly in the aftermath of Partition, which created arbitrary lines dividing communities and ecosystems. The displacement of refugees in the Morichjhapi episode symbolizes the erasure of human lives in the name of territorial integrity and ecological purity. Women characters like *Piya Roy*, an American-born marine biologist of Bengali descent, and Nilima, Kanai's politically engaged aunt, are drawn with depth and agency. Piya, although unfamiliar with the terrain, learns to navigate it through collaboration, particularly with Fokir. Nilima, meanwhile, represents local female leadership. Ghosh offers a nuanced portrayal of women who navigate both natural and social landscapes with resilience, challenging gendered assumptions of passivity or detachment from the environment. The novel foregrounds transcultural interactions, particularly in how foreign characters (Piya and Kanai) relate to local populations. These interactions generate moments of mutual learning and conflict, reflecting the hybrid realities of postcolonial spaces. Ghosh doesn't idealize either side but instead portrays the ethical dilemmas of cross-cultural engagement, especially in the context of science, conservation, and memory. Nature in *The Hungry Tide* is not a backdrop but an active force—unpredictable, generative, and sometimes cruel. The tidal rhythms, monsoon rains, and presence of the Royal Bengal tiger shape human decisions and fates. Ghosh imbues nature with agency, positioning it as both provider and destroyer, reinforcing the need for ecological humility and coexistence.

Ghosh employs a multi-voiced narrative, moving between first-person and third-person perspectives, past and present, to weave together personal, historical, and ecological threads. His storytelling is rich in local myths, oral histories, and vernacular language, creating an immersive cultural texture.

Kanai, a translator and businessman, begins as an outsider—detached and slightly arrogant—but transforms as he reads his late uncle's journal and confronts the lived history of the region. His developing relationship with Piya, whose scientific rationalism is tempered by emotional awakening, forms a central arc in the novel. Fokir, although illiterate, serves as a symbolic counterpoint to the intellectual characters, embodying the intuitive harmony between humans and their environment. His tragic fate ultimately underscores the sacrifices made by the marginalized in the name of progress and preservation.

The Hungry Tide is a masterful exploration of how *climate justice*, historical trauma, and cultural identity converge in a landscape as beautiful as it is brutal. Amitav Ghosh crafts a narrative that is as intellectually rich as it is emotionally resonant, compelling readers to reconsider the binaries between science and spirituality, progress and preservation, outsider and insider. In doing so, the novel not only narrates the story of a place but also becomes a call to ethical environmental engagement in a world where climate change is both a physical and moral crisis.

2. Carpentaria by Alexis Wright.

Alexis Wright's Carpentaria (2006) is a landmark in Indigenous Australian literature that defies generic boundaries and redefines narrative politics. Set in the fictional town of Desperance on the Gulf of Carpentaria in northwest Queensland, the novel delves into the lives of the Aboriginal inhabitants, particularly the powerful figures of Norm Phantom, Angel Day, and Mozzie Fishman. At once mythic, political, and ecocritical, Carpentaria constructs a bold literary topography where Indigenous Law, sacred geographies, and climate consciousness converge. Through its linguistic inventiveness and resistance to linear Western storytelling, the novel becomes a vessel of *climate justice* as seen through Indigenous cosmologies and ontologies.

In Carpentaria, Alexis Wright reconstructs Indigenous cultural memory and oral tradition to depict the struggles of the Waanyi people. The town of Desperance is a symbolic and literal battleground where competing cosmologies—Indigenous and settler-colonial—collide. The narrative addresses environmental exploitation, particularly through the extraction industries that damage the land, and frames this damage not merely as ecological destruction but as a spiritual violation.

The characters, representing archetypes and real-world analogues, serve as conduits through which Wright explores themes of resistance, survival, sovereignty, and environmental stewardship. Through their stories, the novel challenges the economic colonialism of mining companies, the corrupt complicity of government institutions, and the cultural silencing of Aboriginal people.

Unlike Western frameworks that often separate the natural from the cultural, Carpentaria envisions a world where land, spirit, history, and future are entangled in sacred continuities. The novel critiques the capitalist exploitation of natural resources, portraying the mining industry as a desecrating force that violates not only the physical land but also the ancestral Dreaming tracks and spiritual narratives of the Aboriginal peoples.

Wright presents the environment as a sentient entity, alive with sacred resonance. Environmental degradation is not framed as an abstract scientific or global problem but rather as a localized, spiritual catastrophe with profound generational repercussions. In doing so, Carpentaria reorients the discourse of *climate justice* to foreground Indigenous relationships with land, embedding ecological care within cultural and ceremonial practice.

Myth in Carpentaria is not an escape from politics—it is politics. Wright mobilizes Indigenous cosmology and sacred Law as epistemological resistance against colonial structures of knowledge and power. This narrative strategy aligns with Walter Benjamin's notion of "divine violence", whereby Indigenous Law—when activated through myth and ceremony—becomes a transformative and disruptive force.

Mozzie Fishman, a spiritual nomad, embodies this tension between religious activism and cultural reclamation. He journeys along Dreaming tracks to keep spiritual knowledge alive in a world increasingly severed from the land. In contrast to settler law, which seeks to regulate and dominate, Indigenous Law in *Carpentaria* is fluid, organic, and intergenerational, rooted in place and kinship.

Wright subverts Western linear temporality by introducing a Waanyi temporality—cyclical, nonlinear, and deeply symbolic. Time flows like the tides in the Gulf of *Carpentaria*, echoing ancestral rhythms and spiritual recurrence. Past and future are not opposites but layered realities, constantly intersecting in the present. This approach to time challenges the modernist emphasis on progress and historical rupture, presenting instead a temporal continuity that prioritizes responsibility to both ancestors and descendants.

This sacred temporality also allows Wright to frame climate change not as an apocalyptic future but as a current crisis of cultural disconnection, land theft, and environmental violence—unfolding over centuries of colonization.

Wright resists Eurocentric literary conventions through a polyphonic, non-linear narrative that reflects Aboriginal oral traditions. Her "oral narrators" do not simply recount events—they invoke ancestral memory, challenge colonial logic, and engage readers in dialogic ethics. The structure mirrors the ceremonial nature of Indigenous storytelling, which relies on repetition, circularity, and variation.

The narrative is framed by minimal text at the beginning of chapters, mimicking song cycles or Dreaming narratives. These embedded narratives elevate Indigenous epistemologies and position storytelling as an act of sovereignty and reclamation. In *Carpentaria*, to tell a story is to assert existence, to protect the land, and to resist colonial erasure.

This oral tradition is neither monolithic nor nostalgic. Wright's language is richly hybrid, combining Indigenous vernacular, bureaucratic jargon, Biblical allusion, and political satire. This linguistic plurality further enacts decolonial resistance, unsettling readers' expectations and urging them to relearn how to read from an Indigenous perspective.

The characters in the novel showcase ecological and cultural agents.Norm Phantom, the patriarch and fierce land protector, channels ancestral knowledge into pragmatic environmental and political engagement. His spiritual dialogues with the sea and land represent a symbiosis foreign to Western legal paradigms.Mozzie Fishman embodies the sacred nomad, connecting land and spirit through constant movement and ceremony. His character reinforces ecological custodianship as a sacred duty.Angel Day, Norm's estranged wife, is portrayed as a complex, powerful woman whose social isolation reflects the fragmentation of Aboriginal communities under colonial pressure. Yet her interior strength and emotional complexity challenge stereotypes of victimhood.

The ensemble cast demonstrates the resilience and dynamism of Indigenous cultures, with each character contributing to a multifaceted vision of cultural survival and ecological belonging.

Carpentaria is more than a novel; it is a living archive of sacred land, memory, and resistance. Alexis Wright crafts a mythopoeic narrative that unsettles dominant paradigms of law, literature, and environmentalism. In doing so, she offers a uniquely Indigenous articulation of *climate justice*—one that emphasizes relationality, reciprocity, and regeneration over extraction and domination.

Through its sweeping narrative, Indigenous epistemological grounding, and critique of environmental violence, Carpentaria calls upon readers to decolonize not only their politics but also their imaginations. It reaffirms that true *climate justice* is not only about saving the planet but also about restoring stolen lands, honoring ancestral knowledge, and listening to the stories the land still tells.

3. *New York 2140* by Kim Stanley Robinson.

Published in 2017, Kim Stanley Robinson's *New York 2140* emerges as a cornerstone of climate fiction (cli-fi). This genre imagines the social, political, and ecological ramifications of climate change through speculative storytelling. Set in a partially submerged Manhattan following a "second pulse" of sea-level rise, the novel reimagines the city as a post-apocalyptic yet functional archipelago composed of flooded financial districts, habitable skyscrapers, and restructured social systems. Yet, rather than descending into dystopia, New York 2140 resists despair by embracing the possibility of collective action, radical adaptation, and utopian imagination in an age of ecological crisis and capitalist decline.

Robinson, long acclaimed for his environmentally conscious science fiction—particularly the Mars Trilogy—melds his signature eco-socialist ideology with deep research, financial critique, and historical consciousness to construct a narrative that is both an alarm and a blueprint for resistance.

The novel is set in the mid-22nd century, when global warming has triggered two massive sea-level rises, flooding coastal cities and displacing billions. Despite the catastrophic effects of climate change, humanity has adapted. New York survives, transformed into a "SuperVenice" where people traverse canals and inhabit elevated or floating structures.

The narrative centers on residents of the Metropolitan Life Tower, a semi-autonomous vertical community that includes a coder, a hedge fund analyst, a building superintendent, a detective, an internet video star, and a pair of homeless but resourceful children. Their intersecting stories form a mosaic of contemporary concerns: housing, speculation, environmental justice, technological progress, and community resilience. These characters reveal the diverse ways people survive and resist within a system dominated by financial capitalism and haunted by ecological collapse.

Robinson's *New York* 2140 reconfigures the city not as a cautionary ruin but as a site of adaptation. His speculative realism envisions human beings as ingenious survivors, transforming disaster into opportunity. From floating farms to communal housing cooperatives, the city is full of micro-utopias, exemplifying resilience, cooperation, and systemic innovation.

However, Robinson does not idealize suffering. The submerged world is marred by class inequality, ecological displacement, and precarity. While the elite retreat to fortified towers or offshore wealth havens, the majority of residents rely on grassroots solidarity. This vision of disaster utopia underscores a key proposition: that community resilience—rather than techno-fixes or corporate bailouts—will determine the future of climate-stricken societies.

A central axis of the novel is its critique of finance capitalism. The city's real estate is subject to a second speculative bubble, fueled by investors commodifying post-catastrophe urban spaces. Robinson presents financial capital not as an abstract economic mechanism but as a force of environmental devastation and social alienation.

Characters like Franklin Garr, the hedge fund analyst, embody the contradictions of capitalism: he seeks profits in flooded real estate markets while gradually realizing the unsustainability of speculative economies. Through plotlines involving cooperative resistance, rogue investments, and public revolt, the novel imagines horizontal political action—ranging from squatting to mass strikes—as vital to reclaiming urban futures.

Drawing from theorists like Fredric Jameson, Robinson rejects the binary of utopia/dystopia. Instead, *New York 2140* portrays utopia as a process embedded in the everyday acts of care, resistance, and community-building. Whether in the form of neighborhood councils, shared resources, or decentralized media platforms, these emergent structures suggest that another world is possible—not through top-down reform but through collective, situated struggles.

Moreover, the novel's speculative elements engage in cognitive estrangement, a science-fictional technique described by Darko Suvin, which enables readers to view their own socio-ecological systems from a defamiliarized distance. This narrative method not only critiques the present but activates a political imaginary capable of imagining better futures.

Robinson's temporal scope extends beyond the typical climactic event; instead, he emphasizes deep time and long histories of resistance. The novel frequently references historical analogies—Occupy Wall Street, Hurricane Sandy, and the 2008 financial crisis—connecting past and present in a continuum of systemic failure and renewal.

Through a metafictional narrator—sometimes a historian, sometimes the city itself—Robinson offers a meditation on how historical agency is made. Unlike disaster films where salvation comes from heroes or technology, in *New York 2140*, it arises from collective intelligence and democratic action.

Critics have hailed *New York 2140* as a radical climate novel that defies the conventions of dystopian fiction. While many cli-fi texts center on collapse or escape, Robinson offers a vision of pragmatic hope—neither blindly optimistic nor paralyzed by despair.

Some scholars see the novel as part of an emergent ecotopian tradition, alongside Ernest Callenbach's Ecotopia and Octavia Butler's Parable of the Sower, which link environmental sustainability with systemic transformation. Others praise Robinson's emphasis on institutional critique, seeing the novel as a literary intervention into neoliberal hegemony.

Notably, the book has been lauded for foregrounding economic justice as inseparable from *climate justice.* It highlights how environmental risk is unequally distributed, shaped by legacies of racial, geographic, and class-based oppression. The novel's vision of resistance involves reclaiming the commons, rebuilding public infrastructure, and redistributing wealth—a bold program for an ecologically viable future.

In *New York 2140*, Kim Stanley Robinson charts the rising tides of both ocean water and human resistance. Rather than succumbing to the fatalism that often pervades climate discourse, the novel insists that another city, another world, is possible—one forged through solidarity, innovation, and political imagination.

The novel does not offer easy solutions, nor does it romanticize suffering. Instead, it centers the messy, vibrant, collective labor of world-making. In this, *New York 2140* serves as both a mirror of our contemporary crises and a map toward alternative futures—fluid, fragile, but profoundly human.

Storytelling as Resistance: The Role of World Literature in Advancing Climate Justice.

The intersection of *climate justice* and world literature offers a profound and necessary lens through which to reimagine our relationship with the planet, its people, and the systems that shape both. World literature—transcending geographical, linguistic, and cultural boundaries—provides a unique platform for marginalized voices, particularly those most affected by ecological degradation and climate inequity. Through narratives that expose environmental injustices, challenge dominant capitalist ideologies, and envision alternative futures, literature becomes an act of resistance and a space for radical hope.

By engaging with diverse texts—from Indigenous eco-narratives and postcolonial ecologies to speculative climate fiction—readers are invited to confront the uneven distribution of climate burdens and responsibilities. These stories do not merely reflect environmental crises; they actively shape the discourse around justice, agency, and accountability. In a world increasingly defined by ecological uncertainty, world literature stands as both witness and participant, urging collective awareness and transformative action.

Ultimately, the convergence of *climate justice* and literary expression affirms the power of storytelling to heal, connect, and mobilize. It challenges us to listen, to empathize, and, most importantly, to act toward a more just, sustainable, and inclusive planetary future.

CHAPTER 12

Eco-Washed Narratives: Tracing *Greenwashing* in Climate Fiction.

In the context of climate fiction, or "cli-fi," *greenwashing* emerges as a critical theme that exposes the disconnect between environmental rhetoric and actual practices. *Greenwashing* refers to the strategic use of environmental language and imagery by corporations, governments, and other entities to appear environmentally friendly, often while continuing to engage in environmentally harmful behaviour. This deceptive marketing practice is frequently critiqued in climate-related literature, both fiction and non-fiction, which seeks to unveil the mechanisms behind ecological misinformation and corporate irresponsibility. By embedding critiques of *greenwashing* within their narratives, authors shed light on how sustainability is commodified and used to veil ecological damage. The selected works—The Corporation by Joel Bakan, *The End of Nature* by Bill McKibben, *The Sixth Extinction* by Elizabeth Kolbert, and *Flight Behaviour* by Barbara Kingsolver—each offer a nuanced examination of *greenwashing*, illustrating how it affects public consciousness and delays genuine environmental reform.

Greenwashing is a critical theme within climate fiction that refers to the deceptive practice of promoting products, policies, or actions as environmentally friendly when they are, in fact, harmful or insufficient. *Greenwashing* not only distorts public perception of ecological progress but also undermines urgent efforts to address climate change by offering feel-good illusions of sustainability. Literature has become a powerful tool in unmasking these deceptions, critiquing the systemic ways in which corporations, governments, and media outlets manipulate ecological narratives to serve economic interests.

This essay examines how *greenwashing* is represented and critiqued in four major works—Joel Bakan's The Corporation, Bill McKibben's *The End of Nature, Elizabeth Kolbert's The Sixth Extinction, and Claire Watkins' Gold Fame Citrus.* While differing in form and focus, these texts all expose the rhetorical, political, and ethical dimensions of *greenwashing*. They challenge the myth of corporate environmentalism, highlight the dangers of performative sustainability, and call for deeper accountability in environmental stewardship. Through their narratives, these works not only illuminate the mechanisms of *greenwashing* but also provide a literary framework for resisting ecological misinformation and imagining more honest forms of engagement with nature.

1. *The Corporation* by Joel Bakan: *Greenwashing* and Corporate Personhood.

Joel Bakan's 'The Corporation: The Pathological Pursuit of Profit and Power' (2004) offers a penetrating analysis of corporate behaviour, arguing that the legal construct of the corporation, treated as a "person" under the law, exhibits psychopathic tendencies. Central to Bakan's critique is the way corporations utilize *greenwashing* to shield themselves from environmental accountability. Through a series of case studies and interviews with corporate insiders, Bakan reveals how multinational corporations co-opt environmental language to rebrand themselves as eco-conscious while continuing to exploit natural resources unsustainably.

One of the most compelling examples explored in the book is British Petroleum's (BP) "Beyond Petroleum" campaign, which rebranded the company as a forward-thinking leader in clean energy. Bakan (2004) argues that despite these claims, BP remained primarily invested in oil extraction and played a significant role in contributing to greenhouse gas emissions. Such practices exemplify the essence of *greenwashing*: masking destructive behaviour behind a façade of sustainability. The disjunction between corporate marketing and actual practices is not merely misleading but strategically designed to pacify regulatory bodies and the public.

Bakan also critiques the role of public relations firms and sustainability reports in constructing a deceptive image of environmental stewardship. These reports often emphasize minor eco-initiatives—such as office recycling programs—while ignoring large-scale ecological destruction caused by core operations (Bakan, 2004). The result is a manufactured narrative in which corporations appear to be environmental leaders, thereby deflecting attention from their true ecological footprint.

Moreover, The Corporation touches upon the complicity of governmental agencies that often fail to enforce environmental regulations effectively. The deregulation trend of the late 20th and early 21st centuries enabled corporations to write their own rules under the guise of voluntary sustainability. This regulatory capture, as Bakan describes, reinforces *greenwashing* by limiting institutional checks on environmental misconduct.

In essence, Bakan's work is a foundational critique of how corporate structures are inherently aligned against environmental accountability. His exploration of *greenwashing* underscores the urgent need to reform legal and regulatory frameworks to prevent corporations from profiting at the expense of the planet under the pretext of ecological concern.

2. *The End of Nature by Bill McKibben: Greenwashing* and the Loss of Natural Autonomy.

Bill McKibben's *The End of Nature* (1989) is widely regarded as one of the earliest popular environmental texts to communicate the consequences of climate change to a general audience. McKibben presents a sobering thesis: that human activity, particularly fossil fuel consumption and industrial expansion, has altered the Earth's climate so drastically that nature as we once understood it—untouched and autonomous—no longer exists. Within this framework, *greenwashing* is portrayed as a cultural symptom of denial, a means by which society pretends that nature can still be saved through superficial measures while continuing unsustainable habits.

McKibben is especially critical of corporate entities that market eco-friendly products or adopt green branding, while doing little to reduce their overall ecological impact. For example, he critiques how industries promote recycled packaging or energy-saving appliances without addressing the systems of overproduction and hyper-consumption that fuel environmental degradation (McKibben, 1989). He argues that this brand of environmentalism—what we might now label as "green consumerism"—offers a false sense of action and responsibility, distracting the public from the structural changes required to mitigate ecological collapse.

Another crucial element of McKibben's critique is how *greenwashing* fosters complacency. By projecting an image of environmental responsibility, corporations discourage citizens from demanding policy changes or lifestyle shifts. This manipulative optimism, he suggests, undermines meaningful environmental activism, and delays legislative action. McKibben's lamentation of the "end of nature" is thus also a call to reject illusionary fixes and confront the reality of ecological harm with honesty and urgency.

What distinguishes McKibben's perspective is his moral framing: the destruction of nature is not just a technical failure but a spiritual and philosophical one. *Greenwashing*, from this angle, becomes a moral failure—a calculated deception that numbs the collective conscience while the Earth continues to warm.

3. *The Sixth Extinction* by Elizabeth Kolbert: *Greenwashing* in the Age of Mass Extinction.

Elizabeth Kolbert's Pulitzer Prize-winning *The Sixth Extinction: An Unnatural History* (2014) details the ongoing biodiversity crisis caused by human actions, arguing that we are living through the sixth mass extinction event in Earth's history—one driven not by natural forces, but by anthropogenic interference. Kolbert weaves together scientific research, field reports, and historical analysis to show how species loss is accelerating at an alarming rate. Interwoven in this narrative is a critique of how *greenwashing* masks the severity of ecological collapse.

Kolbert highlights how certain industries—particularly agriculture, fossil fuels, and pharmaceuticals—engage in environmentally harmful practices while publicly adopting green marketing strategies. She references examples where multinational companies sponsor conservation programs or wildlife sanctuaries, only to simultaneously degrade the same ecosystems through pollution, deforestation, or habitat encroachment (Kolbert, 2014). These acts of "performative environmentalism" allow companies to gain social capital while continuing to drive extinction-level threats.

A notable insight in Kolbert's work is how *greenwashing* contributes to a false dichotomy: it separates human enterprise from ecological consequence. By pretending that economic growth and environmental preservation can coexist without compromise, corporations promote unsustainable development as benign or even beneficial. Kolbert critiques this neoliberal logic, warning that such narratives lull the public into underestimating the gravity of biodiversity loss.

Moreover, Kolbert draws attention to how *greenwashing* infiltrates the language of policy and international agreements. Terms like "sustainable development" are often co-opted by governments and institutions to justify projects that are minimally compliant or outright destructive. This linguistic manipulation, she argues, undermines conservation efforts by diluting the urgency of scientific findings.

Kolbert's narrative underscores a painful irony: even as we become more aware of ecological decline, we are increasingly exposed to propaganda that insists all is well. *Greenwashing*, therefore, functions as both a smokescreen and a coping mechanism—obscuring truth while enabling continued harm.

3. *Gold Fame Citrus* by Claire Vaye Watkins

Claire Vaye Watkins' *Gold Fame Citrus* (2015) is a haunting and lyrical exploration of ecological collapse set in a drought-ravaged, post-apocalyptic California. The novel explores the environmental consequences of unchecked resource extraction and government inaction, while critiquing the seductive power of *greenwashing* and the cultural mythologies that allow environmental ruin to persist.

At its core, the novel follows Luz and Ray, a young couple squatting in an abandoned celebrity mansion, as they navigate a sun-blasted wasteland known as the "Amargosa Dune Sea"—a constantly shifting mass of sand that has swallowed much of the American West. The novel refrains from pinpointing a single cause for the disaster, instead portraying it as the culmination of systemic ecological neglect and hubris. Watkins uses surreal and poetic prose to depict a world not just ruined by climate change, but spiritually hollowed out by the loss of faith in truth and sustainability.

Greenwashing emerges as a subtle but persistent theme. The government's evacuation campaigns and the false promises of resettlement zones in the East serve as allegories for how institutions package ecological destruction as opportunity. The so-called "Repopulation Bureau" appears as a paternalistic authority that cloaks its manipulative policies in the language of environmental stewardship. As scholar Adeline Johns-Putra argues, "Cli-fi often critiques the way technocratic solutions and corporate rhetoric attempt to reframe climate collapse as manageable rather than systemic" (Johns-Putra, 2016), a critique Watkins seamlessly embeds in her narrative fabric.

Moreover, the cultish figure of Levi Zabriskie, whom Luz and Ray encounter in the Dune Sea, represents a different kind of *greenwashing*: spiritual and ideological. His community mimics the tactics of charismatic eco-leaders who offer hope through self-sufficiency and return to nature, but ultimately reproduce the same power structures they claim to reject. In this sense, Watkins critiques both mainstream and fringe responses to climate disaster, revealing how each can be co-opted by ego, delusion, or control.

The novel also interrogates the language of sustainability and progress. The title itself—Gold Fame Citrus—references California's historical lures: gold for wealth, fame for Hollywood, citrus for agriculture. These motifs echo the state's long tradition of selling dreams while extracting resources, symbolizing the intersection of ecological fantasy and exploitation. As literary critic Heather Houser notes, the novel "exposes the cracks in environmental storytelling, drawing attention to how narrative shapes our engagement with the climate crisis" (Houser, 2020).

In terms of style, Watkins employs speculative realism infused with desert gothic aesthetics. Her fragmented, dreamlike prose destabilizes the reader, echoing the psychological disorientation of her characters. This stylistic choice enhances the novel's central thesis: that collapse is not only physical but perceptual—a breakdown in how people make sense of their relationship with nature, authority, and hope.

Gold Fame Citrus is a significant contribution to contemporary climate fiction for its layered exploration of *ecological grief*, institutional failure, and the seductive illusions of *greenwashing*. Claire Vaye Watkins does not just depict a world ruined by environmental collapse; she critiques the very narratives and ideologies that allow such ruin to be repackaged as manageable or even redemptive. The novel's power lies in its ability to unsettle, challenge, and provoke reflection on the role of both fiction and myth in shaping our ecological future.

Unmasking Illusions: The Literary Reckoning with Greenwashing.

In conclusion, climate fiction plays a pivotal role in exposing the seductive yet dangerous façade of *greenwashing*. By weaving critical insights into compelling narratives, authors challenge readers to question the authenticity of eco-conscious branding and the systemic complicity that enables environmental deception. These stories do not merely critique corporate hypocrisy; they empower readers with the discernment needed to navigate a world saturated with greenwashed promises. In doing so, climate literature becomes a form of resistance—demanding transparency, fostering ecological literacy, and advocating for genuine sustainability over superficial gestures. Through this literary lens, *greenwashing* is revealed not only as a marketing strategy but as a moral and ecological failure—one that must be confronted if we are to build a truly sustainable future.

CONCLUSION

Literature as a Compass in the Climate Crisis Era

The *Anthropocene* epoch has thrust humanity into uncharted territory, where the interconnected fragility of ecosystems and societies demands urgent attention. Literature, an enduring chronicle of human experience, is emerging as a beacon—a space where reflection meets action, and imagination meets resilience. As humanity faces *ecological grief*, climate anxiety, and the turmoil of *polycrisis*, fiction becomes both witness and warning, an invaluable guide through the wreckage.

Far from being passive mirrors, literary works in this transformative era engage actively with the climate crisis. They are instruments of cultural reawakening, shifting from themes of conquest and separation from nature to narratives of collapse and interconnectedness. This evolution captures the essence of climate fiction (Cli-Fi)—a literary movement where ecological reality and imaginative possibility converge to process the *Anthropocene*'s impact on Earth's systems.

Through dystopian worlds and speculative near-present realities, literature provides emotional salves and intellectual provocations, preparing humanity for uncertain futures. Works steeped in themes of *ecological grief*, *deep adaptation*, and *collapsology* offer transformative spaces for readers to engage with vulnerability while envisioning possibilities for change.

In the climate era, literature bears witness to unraveling ecological systems while issuing a clarion call about what may lie ahead. Novels addressing eco-anxiety, *climate refugees*, and existential dread transform abstract data into emotionally tangible experiences. At the same time, these stories serve as warnings, prompting action to counter the inertia of ecological collapse.

Narratives like Margaret Atwood's *Oryx and Crake* and Octavia Butler's *Parable of the Sower* stand as testaments to literature's prophetic power, critiquing escapism and exposing global inaction. The imaginative exploration of *polycrisis*, *Marsification*, and climate-induced displacement foregrounds literature's ability to reflect moral responsibility and urge collective change.

Speculative fiction, particularly Cli-Fi, reimagines survival and coexistence in an era of planetary crisis. By integrating the frameworks of *collapsology* and *deep adaptation*, these narratives challenge survivalist tropes in favor of resilience and regeneration. The apocalypse becomes not an end, but a narrative horizon—a chance to write new stories that honor biodiversity and embrace ecological humility.

The concept of *eutierria* (oneness with Earth) amplifies speculative literature's resistance to human-centered paradigms. These experiments celebrate relationality, acknowledging the agency of the more-than-human world. Cli-Fi thus fosters a cultural shift toward planetary consciousness, urging readers to relinquish control and engage with interconnected resilience.

Literature's intersection with environmental justice

movements underscores its transformative capacity. Voices from the margins—Indigenous narratives, postcolonial perspectives, and stories of *climate refugees*—highlight the unequal burdens of ecological collapse while amplifying calls for reparative justice. By blending empathy with activism, fiction helps readers confront systemic violence and imagine fairer futures.

The narrative shift from anthropocentrism to ecocentrism in global literature deepens the thematic interplay of justice, interconnected struggles, and ecological solidarity. Literature not only illustrates suffering but also inspires liberation, bridging the abstract and the personal to catalyze systemic solutions.

As explored in *The Great Shift,* fiction cannot provide material resources for survival but can transform our mental, emotional, and cultural frameworks. Stories addressing *ecological grief*, climate engineering, and *greenwashing* contribute to sense-making and inspire collective action. Literature's ability to offer solace without sedation, urgency without despair, marks its profound value in the climate crisis.

In a time of converging crises, storytelling evolves as humanity's moral and intellectual compass. Fiction reshapes what it means to be human amidst planetary collapse, guiding readers toward hope, resilience, and reimagination. The stories we tell do not merely reflect the climate emergency—they shape the consciousness needed to confront it.

Can fiction help us survive? It must. Fiction prepares us not only to endure but to dream anew—to imagine transformations that align humanity with the rhythms of a living planet. Let us wield words as tools of resilience and create stories that inspire the systemic renewal our shared future depends on.

REFERENCE

Albrecht, Glenn. *Earth Emotions: New Words for a New World.* Cornell University Press, 2019.

Atwood, Margaret. *Oryx and Crake, The Year of the Flood, MaddAddam.* Anchor Books, 2003–2013.

Atwood, Margaret. *The Year of the Flood.* Nan A. Talese/ Doubleday, 2009.

Bacigalupi, *P. The Windup Girl.* Night Shade Books.2009

Bacigalupi, *P. The Water Knife.* Alfred A. Knopf.2015

Bakan, J. *The Corporation: The Pathological Pursuit of Profit and Power.* Free Press.2004

Bakhtin, Mikhail. *The Dialogic Imagination:* Four Essays. Translated by Caryl Emerson and Michael Holquist, University of Texas Press, 1981.

Baugh, Amanda B. "Environmental Justice in Science Fiction: Exploring Climate, Race, and Belief in Butler's Earthseed Novels." *Journal for the Study of Religion, Nature and Culture,* vol. 11, no. 4, 2017, pp. 456–474.

Bendell, Jem. *Deep Adaptation: A Map for Navigating Climate Tragedy.* Institute for Leadership and Sustainability (IFLAS), 2018.

Bradbury, R. The Martian Chronicles. Doubleday. 1950

Buell, L. *The Future of Environmental Criticism: Environmental Crisis and Literary Imagination.* Blackwell Publishing. 2005

Butler, Octavia. *Parable of the Sower.* Four Walls Eight Windows, 1993.

Cunsolo, A., & Ellis, N. R. (2018). *Ecological grief* as a mental health response to climate change–related loss. *Nature Climate Change,* 8(4), 275–281. https://doi.org/10.1038/s41558-018-0092-2

Cunsolo, A., & Landman, K. Mourning Nature: *Hope at the Heart of Ecological Loss and Grief.* McGill-Queen's University Press. 2017

Diamond, Jared. *Collapse: How Societies Choose to Fail or Succeed. Penguin,* 2005.

Klein, Naomi. *The Shock Doctrine: The Rise of Disaster Capitalism.* Picador, 2007.

Kolbert, Elizabeth. T*he Sixth Extinction: An Unnatural History.* Henry Holt and Company, 2014.

Morin, Edgar. *Homeland Earth: A Manifesto for the New Millennium.* Translated by Sean Kelly and Roger B. H. Savage, Hampton Press, 1999.

Robinson, Kim Stanley. *The Ministry for the Future*. Orbit, 2020.

Saito, Kohei. *Capitalism in the Anthropocene: Ecological Ruin or Ecological Revolution.* Cambridge University Press, 2020.

Tooze, Adam. *The Polycrisis*: A Primer. Verso Books, 2023.

Diggit Magazine.Where is the *Eco-anxiety in Children's Literature?*.

Dimaline, Cherie. *The Marrow Thieves.* Dancing Cat Books, 2017.

Douglas A. Vakoch and Sam Mickey, E*co-Anxiety and Pandemic Distress:* P*sychological Perspectives on Resilience and Interconnectedness,* edited.

Ghosh, A. *The Hungry Tide.* HarperCollins. 2004

Wright, *A. Carpentaria*. Giramondo Publishing. 2006

Robinson, K. S. New York 2140. Orbit. 2017

Nixon, R. *Slow Violence and the Environmentalism* of the Poor. Harvard University Press. 2011

Chakrabarty, D. "The Climate of History: Four Theses." *Critical Inquiry,* 35(2), 197–222. 2009

Heise, U. K. *Sense of Place and Sense of Planet: The Environmental Imagination of the Global.* Oxford University Press. 2008

Trexler, A. *Anthropocene Fictions: The Novel in a Time of Climate Change*. University of Virginia Press. 2015

Whyte, K. *Indigenous Climate Change Studies: Indigenizing Futures, Decolonizing the Anthropocene. English Language Notes*, 55(1-2), 153–162. 2017

Ghosh, Amitav. *Gun Island*. Penguin Random House India, 2019.

Ghosh, Amitav. T*he Great Derangement: Climate Change and the Unthinkable.* University of Chicago Press, 2016.

glennaalbrecht.wordpress.com

Habila, Helon. *Oil on Water*. Hamish Hamilton, 2010.

Hamid, Mohsin. *Exit West.* Riverhead Books, 2017.

Heise, U. *Sense of Place and Sense of Planet: The Environmental Imagination of the Global.* Oxford University Press. 2008

Heise, Ursula K. *Ecocriticism and the Transnational Turn in American Studies*. American Literary History, vol. 20, no. 1-2, 2016, pp. 381–404.

Heller, Peter. *The Dog Stars.* Vintage Books, 2012.

Herd, David, and Anna Pincus, editors. "Refugee Tales: A Literary Project in Solidarity with Refugees and Detainees." *Comma Press,* www.commapress.co.uk. Accessed April 9, 2025.

Houser, Heather. *Infowhelm: Environmental Art and Liter-*

ature in an Age of Data. Columbia University Press, 2020.

IPCC. *Climate Change 2022: Impacts, Adaptation and Vulnerability*. Contribution of Working Group II to the Sixth Assessment Report of the Intergovernmental Panel on Climate Change, Cambridge University Press, 2022.

Jameson, Fredric. *Archaeologies of the Future: The Desire Called Utopia and Other Science Fictions*. Verso, 2005.

Joanna Macy and Chris Johnstone Macy, Joanna, and Chris Johnstone. *Active Hope: How to Face the Mess We're in with Resilience and Creative Power.* New World Library, 2012. (Updated edition: 2022.)

Johns-Putra, Adeline. *Cli-Fi: A Companion*. Peter Lang, 2016.

Kimmerer, Robin Wall. *Braiding Sweetgrass: Indigenous Wisdom, Scientific Knowledge and the Teachings of Plants*. Milkweed Editions, 2013.

Kingsolver, Barbara. *Flight Behaviour.* HarperCollins, 2012.

Kolbert, E. The Sixth Extinction: An Unnatural History. Henry Holt and Company. 2014

Le Guin, U. K. *The Dispossessed*. Harper & Row. 1974

Leopold, A. *A Sand County Almanac*. Oxford University Press. 1949

Louv, Richard. *Last Child in the Woods: Saving Our Children from Nature-Deficit Disorder.* Algonquin Books, 2005.

Mandel, Emily St. John. *Station Eleven.* Knopf, 2014.

McAnany, P. A., & Yoffee, N. (2010). *Questioning Collapse: Human Resilience, Ecological Vulnerability, and the Aftermath of Empire.* Cambridge University Press.

McKibben, B. The End of Nature. Random House. 1989

Morton, T. D*ark Ec ology: For a Logic of Future Coexistence.* Columbia University Press. 2016

Musk, E. Making Humans a Multiplanetary Species. *New Space,* 5(2), 46-61. 2017

NASA . *Mars Overview.* Retrieved from nasa.gov. 2022

Nicholls, P. Review of Mars Crossing. Locus Magazine, 46(3), 22–23. 2001

Nixon, Rob. *Slow Violence and the Environmentalism of the Poor*. Harvard University Press, 2011.

Ojala, M. Hope and climate change: The importance of hope for environmental engagement among young people. Environmental Education Research, 23(4), 624–639. 2017

Panu Pihkala. Eco-Anxiety and Environmental Education. Finnish Society for Environmental Education, 2020.

Panu Pihkala. *Anxiety and the Ecological Crisis: An Analysis of Eco-Anxiety and Climate Anxiety.*

Powers, Richard. *The Overstory*. W. W. Norton & Company, 2018.

Refugee Tales. *Refugee Tales: Volume* I-IV. Edited by David Herd and Anna Pincus, Comma Press, 2016–2021.

Robinson, K. S. *Red Mars*. Bantam Books. 1992

Robinson, Kim Stanley. *The Ministry for the Future.* Orbit, 2020.

Trexler, Adam. *Anthropocene Fictions: The Novel in a Time of Climate Change.* University of Virginia Press, 2015.

Haraway, Donna. *Staying with the Trouble: Making Kin in the Chthulucene.* Duke University Press, 2016.

Chakrabarty, Dipesh. "The Climate of History: Four Theses." *Critical Inquiry,* vol. 35, no. 2, pp. 197–222. 2009

Heise, Ursula K. *Imagining Extinction: The Cultural Meanings of Endangered Species.* University of Chicago Press, 2016.

Britt Wray Wray, Britt. Generation Dread: Finding Purpose in an Age of Climate Crisis. Knopf Canada, 2022.

Sarah Dimick. *Climate Change Literature Course.*

Scranton, R. *Learning to Die in the Anthropocene: Reflections on the End of a Civilization*. City Lights Books. 2015

Servigne, P., & Stevens, R. *How Everything Can Collapse: A Manual for our Times*. Polity Press. 2015

Shelby, Ashley. *South Pole Station*. Picador, 2017.

Steffen, Will, et al. "Planetary Boundaries: Guiding Human

Development on a Changing Planet." *Science*, vol. 347, no. 6223, 2015, DOI:10.1126/science.1259855

Trexler, Adam. *Anthropocene Fictions: The Novel in a Time of Climate Change*. University of Virginia Press, 2015.

United Nations Environment Programme (UNEP). "The *Triple Planetary Crisis*: Forging a New Relationship between People and the Earth." UNEP, www.unep.org. Accessed 8 Apr. 2025.

Watkins, Claire Vaye. *Gold Fame Citrus*. Riverhead Books, 2015.

Weir, A. *The Martian*. Crown Publishing. 2011

Wells, H.G. *The War of the Worlds*. William Heinemann. 1897

Williams, Florence. *The Nature Fix: Why Nature Makes Us Happier, Healthier, and More Creative*. W. W. Norton & Company, 2017.

Williamson, J. "Collision Orbit," *Astounding Science Fiction*. 1942

www.patheos.com

www.psychologytoday.com

Zubrin, R. *The Case for Mars: The Plan to Settle the Red Planet and Why We Must*. Free Press. 1996